THE ROLE OF EYE MOVEMENTS IN PERCEPTUAL PROCESSES

ADVANCES IN PSYCHOLOGY

88

Editors:

G. E. STELMACH

P. A. VROON

NORTH-HOLLAND
AMSTERDAM • LONDON • NEW YORK • TOKYO

THE ROLE OF EYE MOVEMENTS IN PERCEPTUAL PROCESSES

Edited by

Eugene CHEKALUK
Cumberland College of Health Sciences
The University of Sydney
Lidcombe, N.S.W., Australia

Keith LLEWELLYN
Department of Psychology
University of New South Wales
Kensington, N.S.W., Australia

1992

NORTH-HOLLAND
AMSTERDAM • LONDON • NEW YORK • TOKYO

NORTH-HOLLAND
ELSEVIER SCIENCE PUBLISHERS B.V.
Sara Burgerhartstraat 25
P.O. Box 211, 1000 AE Amsterdam, The Netherlands

Library of Congress Cataloging-in-Publication Data

The Role of eye movements in perceptual processes / edited by Eugene Chekaluk, Keith Llewellyn.
p. cm. -- (Advances in psychology ; 88)
Includes bibliographical references and index.
ISBN 13: 978-0-444-89005-4

1. Eye--Movements. 2. Visual perception. I. Chekaluk, Eugene. II. Llewellyn, Keith. III. Series: Advances in psychology (Amsterdam, Netherlands) ; 88.
[DNLM: 1. Eye Movements--physiology. 2. Visual Perception. W1 AD798L v.88 / WW 105 R745]
QP477.5.R65 1992
152.14--dc20
DNLM/DLC
for Library of Congress 92-13138
CIP

Transferred to Digital Printing 2007

PREFACE

It has become a truism that the frozen optical diagram representation of vision is the worst possible picture of the way in which we visually interact with the environment. Even apart from our reaction to moving targets by pursuit movements, our visual behaviour can be said to be characterised by eye movements. We sample the environment in a series of relatively brief fixations, moving from one point to another in the extremely rapid jerks known as saccades. This characteristic of vision raised the question of how our visual world maintains its perceptual stability despite the drastic changes in input associated with these eye movements. This question is, of course, an oversimplification of the problem, as we have to recognise that there may well be differing degrees of stability (and different contributing factors) for the varying properties of the visual world.

One such property is visual direction. In the first chapter, Wayne Hershberger and Scott Jordan look at this question. They represent the modern development that recognises that the two aspects of eye movement - the optomotor control and the associated perceptual effects - may have somewhat different bases. Recognition of this possibility allows for the explanation of what otherwise appear to be quite paradoxical experimental results.

The intermittent nature of visual input via fixations separated by eye movements suggests a number of questions. Helene Intraub takes up one of these in Chapter 2. This is the question of the relationship between the properties of the scene that is actually viewed and the stored memory of that scene. Miguel Garcia-Perez's contribution (Chapter 3) addresses another aspect. As the eye movements take the fixation point from one part of the stimulus to another, the grossly non-uniform stimulus processing features of the basic visual system become highly relevant. The limitation of the effective processing of high spatial frequencies to the central area allows for the operation of quite local stimulus features which may be vital to the perceptual integration of a figure. Thus the nature of figures has something to say about those intriguing phenomena, multistable figures.

The highly foveated nature of our eyes also ensures that we automatically "look at" any object of interest, so that the image of the object falls on the central area. The relationship between interest (or attention) and fixation is so strong that the backward assumption is usually made - an object that the eyes are directed towards is being attended to. In Chapter 4, Geoffrey Underwood and John Everatt examine this "eye-mind" assumption, and point out its limitations. These limitations may have serious implications. The cultural development of one of the editors was irrevocably cut short during high school when his furious music teacher claimed that he had spent 40 minutes staring at a clock when, of course, he had sat enraptured by the music being played, gazing into space. Unfortunately there had been a coincidence between the direction of his eyes and the position of this particular object!

Underwood and Everatt also discuss an issue that has particular force in the area of reading: How do we know where to look next? An obvious response to this question is that we probably jump from one fixation point to the next in steps of standard size. It seems, however, that the jumps are guided by a variety of features of the written material itself. The fact that we can process the meaning of a set of marks on a page quite rapidly despite the brevity of the fixation pauses, the highly restricted area of high acuity and the size of the jumps from one fixation point to the next has intrigued researchers for years. Perhaps the fact that we are simultaneously processing information contributing to the control of the jumps is even more amazing.

A vital question in the area of reading is how we can have such a rapid sequence of inputs from the series of fixations and not end up with a confusing composite. The question of how the "slate is wiped clean" between fixations is only one of a set of questions that arise whenever the implications of saccades are considered. From a saccadic omission point of view the question is why we do not see a horrible blur or "gray-out" during the saccade. From a saccadic suppression point of view the question is that of the nature and explanation of the change in visual sensitivity which actually starts before the onset of the saccade. Eugene Chekaluk and Keith Llewellyn examine the effects of visual masking in these situations in Chapter 5.

Our freedom from the perceptual consequences of the drastic input changes produced by saccades is normally taken for granted, but, as noted above, any loss of this freedom could have serious implications for our visual performance. Breitmeyer has invoked the features of transient and sustained channels to account for our freedom from persistence of one fixated scene into the following fixation(s). Bill Lovegrove (Chapter 6) reports on research looking at the relationship between deficits in this transient system and specific reading disability. The essential neuroanatomical and physiological underpinnings of eye movements are further examined in Chapter 7 by John Morley who looks at the specific case of vergence.

This book ends at the beginning. Earlier we mentioned the amazing visual/perceptual/cognitive performance that reading represents. For an adult, however, reading is, after all, a very over-learned skill. In our last chapter (Chapter 8) Sandra Shea draws our attention to the enormously complex motor control problems normal visual behaviour poses to an infant, who is also handicapped by having a system that is changing basic characteristics as he or she develops and grows. She points out that one of the puzzles is just how infants can manage so well. Other puzzles include that of the form of the developmental patterns that occur. Shea also deals with the practicalities of infant visual research, the limitations and applications of particular techniques, and the unfortunate characteristics of the subjects. R.H. Day once described infants as being more interesting subjects than white rats, but the editors are not convinced that we should exchange our standard laboratory animals - undergraduate students - for infants, no matter how scientifically rewarding this might be.

Eugene Chekaluk Keith R. Llewellyn

Sydney, Australia 1991.

ACKNOWLEDGEMENTS

The editors would like to thank Associate Professor John Sheppard/ Mr Ian Robertson and Professor Barbara Gillam, who as respective Heads of Department in which the editors work were supportive of the project and gave all possible assistance.

Thanks are also due to Lily Lee, My Phung Luu, Wendy Manuel and Karen Taylor, WPMOs of the Cumberland College of Health Sciences, University of Sydney who all, at various stages, typed the chapter manuscripts, often to quite unreasonable deadlines.

Special thanks are due to My Phung Luu who formatted the book to the publisher's specifications.

Finally, thanks are due to Dr. K. Michielsen and Ms. Alison Carter, of North-Holland who helped us put the book together.

Eugene Chekaluk Keith Llewellyn

Sydney
Australia
December1991

CONTENTS

CONTRIBUTORS (in alphabetical order)

Eugene Chekaluk, Cumberland College of Health Sciences, The University of Sydney, P.O. Box 170, Lidcombe, N.S.W. 2141 Australia e-mail: BH_CHEKALUK@coco.cchs.su.oz.au

John Everatt, Department of Psychology, University of Nottingham, Nottingham, NG7 2RD England

Miguel Angel Garcia-Perez, Universidad Complutense, Laboratorio de Vision, Facultad de Psicologia, Campus de Somosaguas 28023 Madrid, Spain
e-mail: LABVI02@EMDUCMS1

Wayne A. Hershberger, Department of Psychology, Northern Illinois University, DeKalb, Illinois 60115, U.S.A.
e-mail: TJ0WAH@NIU

Helene Intraub, Department of Psychology, University of Delaware, 220 Wolf Hall, Newark, Delaware 19716, U.S.A.
e-mail:AOB15336@UDELVM

J. Scott Jordan, Department of Psychology, Northern Illinois University, DeKalb, Illinois 60115, U.S.A.

Keith R. Llewellyn, School of Psychology, University of N.S.W., P.O. Box 1, Kensington, N.S.W., 2033 Australia

William Lovegrove, Department of Psychology, University of Wollongong, Wollongong, N.S.W. 2500 Australia
e-mail:w.lovegrove@uow.edu.au

John Morley, School of Physiology and Pharmacology, University of N.S.W., P.O. Box 1, Kensington, N.S.W., 2033 Australia

Sandra Shea, School of Medicine, Southern Illinois University at Carbondale, Carbondale, Illinois 62901, U.S.A.
e-mail:GA3655@SIUCVMB

Geoffrey Underwood, Department of Psychology, University of Nottingham, Nottingham NG7 2RD England
e-mail: GJU@PSYC.NOTT.AC.UK

The Role of Eye Movements in Perceptual Processes
E. Chekaluk and K.R. Llewellyn (Editors)

VISUAL DIRECTION CONSTANCY: PERCEIVING THE VISUAL DIRECTION OF PERISACCADIC FLASHES

WAYNE A. HERSHBERGER and J. SCOTT JORDAN

1. Introduction

This chapter is concerned with the perceptual phenomenon known as visual direction constancy (Shebilske, 1977), particularly as it relates to saccadic movements of the eyes. The current experimental literature on this topic is congested with contradictory experimental findings. Although the conflicting findings are all interrelated, they may be parsed into three distinct paradoxes. In this chapter, we will attempt to show how all three of these paradoxes may be resolved parsimoniously on the basis of a single coherent theoretical account. This theoretical account involves an elaboration of a pair of hypotheses advanced earlier (Hershberger, 1987) as adjuncts to a theoretical model of the saccadic oculomotor system developed by Robinson (1975, 1981, 1986). Robinson's closed-loop model controls eye orientation and utilizes, as do all servo systems, two separate indices of the variable being controlled: a reference signal and a feedback signal. The neural feedback signal in Robinson's model is what von Holst and Mittelstaedt (1950) called an efference copy. The neural reference signal is what we are wont to call an afference copy (Hershberger, 1976, 1987, Hershberger & Misceo, 1983). Our thesis is that saccadic eye movements depend upon both neural copies whereas psychophysical judgments of visual direction depend only upon the afference copy. The former hypothesis is called the sum-of- errors hypothesis, for reasons given below. The latter is called the afference copy hypothesis.

After first identifying the three paradoxes, we will begin the chapter with a formal analysis of the *visual perception of direction*, conducted with a view to identifying the role that the oculomotor system must play in the *constancy* of visual direction. Then, we will show how all three empirical paradoxes may be resolved by a modified version of Robinson's (1975) oculomotor theory.

2. Three Paradoxes

Matin versus Halett and Lightstone: Seeing versus looking

Matin and his associates (Matin, 1972, 1982; Matin & Pearce, 1965, Matin, Pearce, Matin, & Kibler, 1966; Matin, Matin, & Pearce 1969; Matin, Matin,

& Pola, 1970) have found that the apparent visual direction of a point light source flashed very briefly (e.g., 2 ms) during a saccade in total darkness is illusory; yet, Hallett and Lightstone (1976a, 1976b) have found that if one intends to look at such a flash of light, a reasonably accurate refixation saccade occurs in due course. The paradox is that one looks for the flash in one direction while professing to see it located in another. Similarly, Ono and Nakamizo (1977) have found that a saccade made during a change of fixation to a target at a different distance is directed toward the target's actual rather than perceived location.

Hallett and Lightston Versus Honda: An apparent failure to replicate

Honda's (1989) recent failure to replicate the results of Hallett and Lightstone seems to offer a way out of the aforementioned dilemma, but that would leave the findings of Hansen and Skavenski (1977) unaccounted for.

Their subjects, using a hammer to strike targets briefly illuminated during saccadic eye movements in the dark, were very accurate, implying veridical perception of the target locations. Hence, rather than resolve the problem, Honda's findings simply exacerbate it. That is, if the experimental findings of Honda, on the one hand, and Hansen and Skavenski, on the other, are both to be believed, how then does one account for the findings of Hallett and Lightstone (or Ono & Nakamizo) which agree with those of Hansen and Skavenski but not Honda?

The Mays and Sparks Paradox: Spatiotopic versus retinotopic co-ordinates

A third, and equally disconcerting, paradox concerns the spatial coding of saccadic eye movements in primates: Whereas Mays and Sparks (1980b) have found evidence suggesting that saccades are coded in terms of a spatiotopic frame of reference, the activity of individual neurons in saccadic control centers of the brain seem to imply that saccades are coded in terms of a retinotopic frame of reference (Robinson, 1986).

In this chapter, we will attempt to show how all three of these paradoxes may be resolved parsimoniously on the basis of a single coherent theoretical account that incorporates the two hypotheses mentioned above as adjuncts to Robinson's closed-loop model of the oculomotor system. Understandably, we will be turning our attention to a careful consideration of Robinson's model. But first, some general introductory remarks about the perception of visual direction.

3. Visual Direction Constancy

Typically, the visible environment does not appear to move when we rotate our eyes, even though eye rotations sweep our retinas across the optic array. That is, in spite of the motion of the proximal stimulus across the retina, the distal environment appears to be undisturbed or to remain stable. The distal environment is said to "remain stable," as if it were in fact motionless. But, of course, our terrestrial environment is not motionless. It is moving at a very high velocity through the solar system as the earth orbits the sun and spins on its axis.

Stability is in the eye of the beholder. Or, more exactly, motion is always relative to a frame of reference. When we take our environment to be motionless we are in fact imposing a particular frame of reference, namely, the surface of the earth.

To say that the distal environment remains stable in spite of eye movements is to say that the visible environment does not appear to move relative to the surface of the earth when we shift our gaze. The perceptual phenomenon known as visual space constancy merely reflects our constant utitlization of the surface of the earth as an allocentric frame of reference; it does not imply that the earth is immobile. Our utilization of the surface of the earth as a perceptual frame of reference is, at once, eminently reasonable and difficult to understand. The practical utility of using such an allocentric frame of reference may be obvious, but how we manage to do so it not.

An Example

In order to illustrate the problem as simply and precisely as possible, consider the following example. Imagine a slowly panning television camera situated at the center of a circular railroad track filled with a long train chasing its tail. As the camera and train each more relative to the earth, the camera's lens sweeps across the optic array by an amount equal to the algebraic sum of the respective motions of the two devices (camera and train), each relative to the earth. If one watches this scene unfold on the screen of a television set (i.e., receiver) it is relatively easy to tell when the train and camera are moving relative to each other; the features comprising the image will move across the screen (here the frame of the television screen serves as the immediate frame of reference). Further, if the train and camera are moving relative to each other, it is apparent that at least one is moving relative to the surface of the earth, but which one is moving may be difficult to determine. Indeed, sometimes it will be impossible to determine which one(s) is(are) moving relative to the earth, and by how much. That is, when the train alone is visible on the television screen, the movements of the images provide absolutely no information about whether or not both devices (train and camera) are moving relative to the earth. By comparison, if the earth itself is visible on the screen, then the respective motions of the two devices are directly presented in the optic array (Gibson, 1966, 1979). However, if this allocentric perceptual frame of reference is not represented on the screen in any form, how could it be imposed upon the optic array so that one could see the televised motion(s) relative to the surface of the earth?

This, essentially is the question one ultimately addresses when considering the nature of the visual phenomenon known as space constancy: seeing the furnishings of the visible environment as being positioned relative to the surface of the earth, even when this surface of support is invisible, and we ourselves are moving across it. Note that the question is independent of the directness (Gibson, 1966, 1979; Turvey & Carello, 1986) or the indirectness (Ullman, 1980; Wallach, 1987) with which optical information is detected. The question does not concern the nature of the optical information itself but, rather, the perceptual framework in which the optical information is depicted.

Continuing the example, suppose that the television receiver projects its images upon a portion of a surrounding cycloramic screen extending a full 360 degrees, and that this projector rotates relative to the earth exactly as does the panning television camera so that the projected image moves across the screen to the same degree that the camera pans across the train. Now the motions of both the camera and the train are represented (imaged) relative to their common surface of support (i.e., the earth). Further, if the directions in which the camera and projector face are perfectly aligned, then the position of the televised image on the screen will faithfully represent the track position of that particular portion of the train. More exactly, the angular orientation of the televised portion of the train, from the camera's point of view, will be faithfully represented by the angular orientation of the image, from the projector's point of view. In other words, visual direction constancy will obtain. For all panning motions of the camera and all motions of the train, the visual direction of the image on the screen will correspond exactly to the visual direction of that portion of the track occupied by the televised portion of the train.

Further, if the tripod supporting the camera should pivot on its base, and the camera pivots on the tripod, visual direction constancy can still be preserved if the projector's tripod is pivoted on its base by an identical amount. That is, if the camera's immediate frame of reference (the tripod) itself rotates relative to the earth, then the rotation of this *intermediate frame of reference* must be mimicked by a corresponding shift in the projector's intermediate frame of reference, if visual direction is to be preserved. Furthermore, if the base of the camera's tripod moves relative to the earth, then the base of the projector's tripod would need to be moved likewise. In general, as long as the projector shifts its intermediate frames of reference relative to the earth in the same way that the camera shifts its intermediate frames of reference relative to the earth, then space constancy will be preserved in the visual (televised) projections. Consequently, we can see that visual space constancy is essentially a matter of properly registering, or taking into account, the relative orientations of all the intermediate frames of reference which link an optical sensor (eye or camera) to the surface of the earth (Epstein 1973, MacKay, 1962).

Before leaving this example, a caveat is in order: We are not suggesting that a projection process is involved in the mediation of visual direction constancy. The projector and the projected images on the cycloramic screen were used merely to illustrate the particular shifts among the intermediate frames of reference linking the camera to the earth which had to be registered in order for the camera to register visual direction relative to a constant frame of reference-earth. The camera will register visual direction relative to the earth providing it registers this extra-optical information along with its optical information. This perceptual constancy does not imply a need to project the information.

4. Visual Frames of Reference

In human vision, the optic array is linked to the earth through three intermediate spatial frames of reference. Beginning with the pencils of light entering the eye, the immediate frame of reference comprises the coordinates of the retinal mosaic (retinocentric or retinotopic coordinates). In turn, the immediate frame of reference for the eye comprises the coordinates of the ocular orbits in the head (egocentric or spatiotopic coordinates). The immediate frame of reference for the head comprises the inertial coordinates of the vestibular apparatus in the bony labyrinth of the inner ear; these inertial coordinates approximate those of the earth (allocentric or environmental coordinates).

When the head is mobile, reflexive counter-rotations of the eyes invariably accompany rotations of the head (the vestibulo-ocular reflex). That is, as the head rotates relative to the earth, the eyes counter-rotate relative to the head, so that the retinal image tends to remain retinotopically stable. In effect, the spatiotopic frame of reference shifts in opposite directions relative to each of the other two sets of coordinates (retinocentric and allocentric). Although this simultaneous shift of the spatiotopic frame of reference relative to the other two has valuable practical consequences, it also confounds the two shifts, making it difficult to asses the respective features of each. Consequently, subjects in experimental investigations of visual-direction commonly have their heads immobilized by biting rigidly mounted dental impressions so that only one type of shift is possible, namely, shifts between the retinocentric coordinates of the eye and the egocentric coordinates of the head. Since we call these shifts eye movements, it is apparent that our analysis of the perceptual phenomenon known as visual direction constancy will necessarily entail a consideration of the nature of the oculomotor control system that rotates the eyes in their orbits.

5. The Oculomotor System

The Oculomotor Plant

The eye moves in its orbit essentially as ball in a socket, rotating about a central point located about 13.5mm behind the cornea (Park & Park, 1933; however, cf. Enright 1984). Although the eye has no fixed axes of rotation, its rotations may be represented as comprising three rotational components about three orthogonal axes (Helmholtz, 1867/1962), one being the line of sight (cyclotorision). The other two axes may be modelled either as Cartesian coordinates, one horizontal and one vertical (e.g., Helmholtz's system), or as polar coordinates in which movements of the eye are represented in terms of angles of eccentricity and meridional direction (e.g, Listing's system).

The orientation of the eye in its orbit is determined by six extraocular muscles, comprising three antagonistic pairs: (a) a vertical pair--the superior and inferior recti, (b) a horizontal pair--the lateral and medial recti, and (c) a torsional pair--the superior and inferior obliques. Although the primary actions of these three pairs of extraocular muscles are vertical, horizontal, and torsional rotations, respectively, each pair of muscles contributes to rotations about more than one axis. Consequently rotations of the eye which are purely vertical or purely horizontal generally involve a synergistic coupling of the effects of all four recti. For instance, a purely vertical rotation of the eye from the primary position ("straight ahead") involves a large component from the horizontal pair of rectus muscles. However, horizontal rotations of the eye from its primary position are almost exclusively the primary action of the horizontal pair, the lateral and medial recti (Boeder, 1961). The extraocular muscles, like other voluntary muscles, comprise two types of fibers, large diameter, phasic fibers that twitch in an all or none fashion in response to neural stimulation, and small diameter, tonic fibers which contract with a force that varies with the frequency of the neural stimulus. Both types of fibers participate synergistically in the control of both steady fixation and dynamic rotation of the eyes (Collins, 1975).

Human extraocular muscles also incorporate intrafusal fibers, or stretch receptors (Cooper & Daniel, 1949). However, the stretch reflex which these receptors mediate in the skeletal musculature (Matthews, 1972) has not been detected in the primate oculomotor system. For instance, Keller and Robinson (1971) monitored the activity of 38 motor neurons in the abducens nuclei of 3 Rhesus monkeys and were unable to find any reflexive activity when they stretched their subjects' lateral rectus muscles. Since their single-cell recording techniques probably monitored only the larger phasic fibers, the tonic fibers may evince a stretch reflex (as some research with rabbits and cats suggests: Baichenko, Matyushkin & Suvorov, 1967). However, if there is such a "tonic reflex" in primates, its utility remains obscure, because there appears to be no type of eye motion in which the phasic fibers do not participate fully.

Although the extraocular muscle may not itself exhibit a stretch reflex, stretching an extraocular muscle does have remote neural effects in diverse parts of the brain, including various layers of the superior colliculus, the cerebellum, the visual cortex, and the frontal eye fields (Howard, 1982). However, the functional significance of this muscle spindle afference is not yet clear.

Neural Control of Eye Movements

The extraocular muscles are innervated by the neurons of three brainstem (oculomotor) nuclei: the abducens nucleus (lateral rectus), the trochlear nucleus (superior oblique), and the oculomotor nucleus (superior, inferior, and medial recti, and interior oblique). Research with monkeys has repeatedly shown that oculomotor neurons (a) have tonic firing rates which are linearly related to eye orientation during steady fixation, with different cells having different ranges of activity, (b) increase or decrease their firing rates with corresponding changes of eye velocity during pursuit eye movements, and (c) evince high firing rates (saturation burst rates) during saccades in the on-direction, or inhibition during saccades in the off-direction, with the burst, or inhibition, beginning and ending about 10 ms before the saccade begins and ends (e.g., Fuchs & Luschei, 1970; Robinson, 1970; Schiller, 1970).

These motor neurons in the oculomotor nuclei serve as the final common path for several different supranuclear control systems (or subsystems), each comprising a semi-independent component of the oculomotor system. These supranuclear components are traditionally classified in terms of three factors considered jointly, (a) the type of eye movement they control (fast or slow), (b) the type of disturbance which drives those eye movements (visual, vestibular, or other), and (c) the type of oculomotor intention involved, including looking at something (fixation or pursuit), looking at something else (saccades), or merely pointing the eye (as in the dark).

The semi-independent components comprising the oculomotor system operate in concert to mediate complex effects. For instance, optokinetic nystagmus is a reflexive alternation of fast and slow counter-rotations of the eyes driven by a drifting retinal image (i.e., a visual disturbance); the slow component reduces retinal slip, and the fast component returns the fovea to a portion of the image nearer the median plane of the head, so that the slow phase may continue (Raphan, Cohen, & Matsuo, 1977). The reflex operates whether or not the eye is simultaneously pursuing a target (looking at something in the drifting image) or merely staring passively into space; in the former case, however, the slow phase of the reflex includes a second component which reduces the retinal eccentricity of the pursued target (Dichgans, 1977).

Such a detailed consideration of all the various subsystems comprising the oculomotor system is beyond the scope of this chapter. For our present purposes, what is important about these subsystems, is that they all share a final common path, the motor neurons in the oculomotor nuclei of the brain stem. That is, the semi-independent components do not have independent access to the extraocular muscles but, rather, control eye orientation only by controlling the output of a common set of motoneurons. Hence, eye orientation, however it is being controlled, is always neurally represented by a common set of neural signals. In fact, these neural signals appear to provide the principal means by which the brain registers eye orientation for the purpose of maintaining constancy in the perception of visual direction. Because these signals are *corollary discharges* of efferent cells (Sperry, 1950), they are generally called *efference copies* (von Holst & Mittelstaedt, 1950).

The muscle spindles (stretch receptors) in the extraocular muscles provide an alternate means of registering eye orientation (Skavenski, 1972; Steinbach, 1986) but the weight of the evidence appears to be overwhelmingly in favor of the efference copy as the principal index of eye orientation (e.g., Skavenski, Haddad, & Steinman, 1972). For example, pressing on the canthus of the eye with one's finger does not shift the apparent location of an afterimage but does shift the apparent visual direction of a fixated object, and in a direction reflecting the increased muscular innervation required to maintain fixation (Bridgeman & Delgado, 1984; Stark & Bridgeman, 1983). Conversely, saccadic movements of the eye will displace the apparent visual direction of negative afterimages in the direction of the eye movement while leaving the location of visual targets undisturbed. (For an historical review of this age-old issue see Grusser, 1986.)

6. Robinson's Model of the Saccadic Control System.

The foundation of our thesis is Robinson's model of the oculomotor control system, particularly the component or subsystem that controls saccadic eye movements (Robinson, 1975, 1981, 1986). Robinson's control model is most readily summarized in the form a flow diagram, which both identifies the relevant variables and shows how they are related to each other causally (i.e., by transfer functions). Figure 1B is a simplified version of Robinson's model that is comparable to a diagram Sparks and Mays (1983a) have used to illustrate the implications of their own research. This simplified version is particularly useful for our present purposes which include our intent to resolve a paradox posed by the experimental findings of Mays and Sparks (1980b). Figure 1A illustrates an earlier version of Robinson's model (Robinson, 1973) in which saccadic eye movements are driven by the retinal eccentricity of the target (i.e., an error signal expressed in retinal coordinates is nulled by the saccade). The retinal eccentricity of the target (RE) is a function of two variables acting jointly, the orientation of the target relative to the head (T/H) and the orientation of the eyes relative to the head (E/H); specifically, the error signal equals T/H minus E/H. This earlier version controls the position of the image on the retina without controlling the orientation of the eyes in the head (the orientation of the eyes in the head is reflexively determined by the location of the target). However, our ability to voluntarily point our eyes in the dark demonstrates that the oculomotor system can control the orientation of the eyes, and implies, further, that the system may actually control the orientation of the image on the retina by controlling the orientation of the eyes in the head. Indeed, this is the way in which Robinson's current model (Figure 1B) operates.

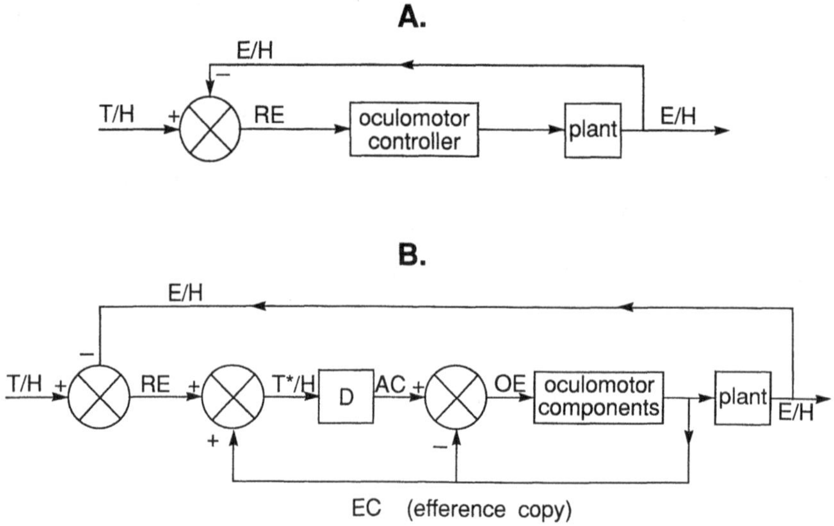

Fig. 1. Simplified versions of Robinson's early retinocentric (A) and current egocentric (B) models of the saccadic system. T/H, target position relative to the head; E/H, orientation of the eyes in the head; RE, retinal error (retinal eccentricity of the target's image) D, delay; T*/H, neural estimate of target position relative to the head; EC (efference copy), sensed orientation of the eyes in the head; AC (afference copy), neural representation of intended eye orientation relative to the head; OE (oculomotor error), the difference between the intended and sensed orientation of the eyes in the head (AC-EC).

Supporting Evidence

The experimental data that provides, perhaps, the best evidence in support of Robinson's current model have been those reported by Mays and Sparks (1980b). Mays and Sparks, investigating saccadic eye movements in rhesus monkeys, used electrical stimulation of the superior colliculus to move the eyes of a monkey just before he began a saccade to a spot of light flashed previously in the dark (i.e., flashed just before the electrical stimulation). Despite this electrode-induced perturbation, and the fact that the flashed target was no longer visible, the monkey's subsequent saccade brought his gaze to the target location, something clearly impossible had the movement been determined solely by retinal information.

Clearly, the oculomotor system controls eye position, rather than eye movements. That is, eye movements are directly driven by an oculomotor error signal rather than a retinal error signal. According to Robinson's current model the muscular innervation driving the eye from one position to another depends upon the difference between two neural signals, each representing eye position, one sensing the position of the eyes (efference copy) and the other specifying the intended position of the eyes. Because the latter signal specifies the intended sensation, it too is a type of copy, namely, a prototype specifying what the sensed eye position is to be (i.e., an *afference copy*; Hershberger, 1976, 1987; Hershberger & Misceo, 1983). Further, because such prototypic signals (also known as reference values, set points, sollwerts, and the like) are routinely mistaken for efference copies (e.g., Gallistel, 1980; Hinde, 1970) we will use the expression afference copy here so as to clearly distinguish between the two different meanings.

In Robinson's model, intended eye orientation is an afference copy (AC), as prototype specifying the intended value of the controlled variable. The controlled variable is eye orientation, sensed by means of an efference copy (EC), a facsimile of the value of the variable being controlled. The algebraic difference between these two variables (AC-EC) comprises oculomotor error (labeled OE in Figure 1B); it is this error signal that the oculomotor feedback loop (on the lower right) nulls.

The oculomotor feedback loop of the model has been well documented; the neural components have been identified in considerable detail (e.g., see Figure 3, Robinson, 1981). However, the neurological substrate for the loop on the lower left is much less definite. The loop on the lower left is concerned with the *estimation* of the orientation of saccadic targets relative to the head (T*/H): Since RE equals T/H minus E/H, and EC is an estimate of E/H, RE plus EC would be an estimate of the targets location relative to the head (T*/H). This estimate serves to specify the intended eye orientation (AC) for that target. That is, according to the model summarized in Figure 1B, the intended orientation of the eyes (AC) is periodically reset (after a delay) equal to the estimated current orientation of the visual target relative to the head (T/H*), with this estimated orientation being the sum of two angles, the retinal eccentricity of the target's image (RE = T/H - E/H), and the angle represented by the efference copy (EC), estimating the orientation of the eyes in the head. Specifically, T/H* - RE + EC. Thus, T/H* is an accurate estimate of T/H to the degree that EC is an accurate estimate of E/H.

Locating Saccadic Targets

Two brain centers are intimately involved with the visual mediation of saccadic shifts of intended eye orientation: the superior colliculus, and the frontal eye fields of the cortex. Only when both of these areas are destroyed is the ability to make visually organized saccades permanently lost (Schiller, True & Conway, 1979). Consequently, researchers have looked in these two areas for neurons that code visual targets in terms of a spatiotopic frame of reference (i.e., T/H*). That is, since the coordinates of the head serve as the frame of reference for the value of AC (intended eye orientation relative to the head) it was assumed that cells could be found in the superior colliculus and the frontal eye fields that code the position of visible targets in terms of that same frame of reference, the coordinates of the head. Although such cells have been found in the thalamus (Schlag & Schlag-Rey, 1983) they have not been found in either the superior colliculus or the frontal eye fields of primates. [Research findings with cats are controversial; some investigators (e.g., Crommelinck, Guitton, & Roucoux, 1977) claim to have found spatiotopic collicular cells in cats, but others have not (Harris, 1980).]

In the frontal eye fields, there are cells that discharge according to the retinal location of a stimulus, according to the direction and amplitude of the most recent eye movement, and according to the metrics of the next visually guided saccade (Bruce & Goldberg, 1981).

> In a two jump experiment, cells discharge not according to the retinotopic target location or the spatial target location, but rather according to the eye movement needed to acquire the target. There is no static map of the world, just a map of saccades. A given retinal

> stimulus can evoke any saccade, given the proper antecedent eye movement. The spatial map is only a virtual map, linked to the motor output by the recent eye movement (Goldberg, 1983, p.21).

In the colliculus, we see a similar picture. In double-saccade trials, in which the target for neither saccade was visible at the time of the saccade, Mays and Sparks (1980a) found that there are collicular cells that discharge according to the retinal location of a stimulus, and according to the metrics of the next visually guided saccade. The movement-coded cells are of two types: one type (found in the deep layer of the colliculus) discharges briefly, 20 ms prior to a saccade; the other type ("quasi-visual cells," found in the intermediate layer) discharges continuously for many milliseconds, beginning well before (e.g., 150 ms) the saccade and ending as the saccade is completed. A quasi-visual cell which fires in response to a target flashed 10 degrees to the left of the monkey's line of sight will also fire when the monkey prepares to look to a remembered invisible target located 10 degrees to the left of the monkey's line of sight. What is common to these two cases is the impending eye movement. Therefore, Sparks and Mays (1983a) suggest that the quasi-visual cells code the oculomotor error, OE, for the impending eye movement. (Wurtz & Goldberg, 1971, made a similar observation a decade earlier.) However, they reported finding no cells in the monkey superior colliculus encoding the position of targets in head coordinates (Sparks & Mays, 1983a).

7. The Paradox of Mays and Sparks Resolved

The findings of Mays and Spark seem paradoxical. On the one hand, some of their findings imply that the AC for the oculomotor control loop is expressed in head coordinates (Mays & Sparks, 1980b), which means that the locations of saccadic targets must also be expressed in head coordinates. On the other hand, when they have looked into the superior colliculus (a saccadic command center) for cells that encode target locations in head coordinates, they report having found none (Sparks & Mays, 1983a).

The Sum-of-Errors Hypothesis

Ironically, the means of resolving this paradox are also provided by the findings of Mays and Sparks. They found two types of error signal coded by two types of cells in the superior colliculus: visual cells, and quasi-visual cells. The visual cells in the superficial layer topographically code the retinal eccentricity (RE) of visual targets. And, the quasi-visual cells in the intermediate level topographically code the value of the oculomotor error signal (OE), the signal that will drive the impending saccade after the customary saccadic latency. The sum of these two error signals provides a basis for specifying saccadic targets in the coordinates of the head--using the current value of AC as an integral base. The oculomotor system already "knows" the values of AC currently in effect, so all that it needs from a saccadic control center, in order to update or correct the current value of AC for a new target, is the sum of two error signals: (a) the retinal eccentricity of the target's image (RE), and (b) the oculomotor error signal (OE) at the time the visual target was presented.

This sum-of-errors hypothesis is illustrated in Figure 2 and incorporated into a diagram of Robinson's model in Figure 3. Figure 3 is virtually equivalent to Figure 1B. Note in Figure 1B that OE equals AC minus EC; therefore, EC equals AC minus OE. In Figure 3, EC is factored into its two components (+AC and -OE, or more exactly +AC* and -OE*) which are added in at two separate summing points. Figure 3 essentially "rephrases" Robinson's model in order (a) to incorporate the sum-of-errors hypothesis, and (b) to illustrate where the visual cells and the quasi-visual cells of the superior colliculus appear to fit into the picture.

There is, however, a slight but significant difference between Figures 1B and Figure 3. In Figure 3, two oculomotor error signals are calculated: OE and OE*. The former (OE = AC-EC) is the familiar variety; it is the error signal extant in the oculomotor feedback loop, the one that drives the saccade required to achieve the *present* reference value, AC. The latter oculomotor error signal (OE* = AC*-EC) is an estimate of the metrics of the *next* saccade, the one required to shift the eye to a *new* target orientation, AC*. The theory implicit in Figure 3 differs from Robinson's to the degree that AC* and OE* are substitued for AC and OE. This substitution converts the target-locating process (mediated principally by the lower left loop in Figure 3) into an error nulling process.

The lower left loop in Figure 3 is a feedback loop that will continually null the sum-of-errors signal, SE. The loop integrates the sum-of-errors signal, thereby incrementing or decrementing the value of AC* until the value of OE* is driven into correspondence with RE. When the sum-of-errors is zero, the value of AC* equals the orientation of the target, expressed in head coordinates. Thus, the lower left loop locates saccadic targets in head coordinates. Consider the following example. (In our examples, we will indicate an error signal's polarity in terms of the eye movement it calls for, with clockwise being positive; and, we will ignore measurement error in our examples in order to keep them simple.)

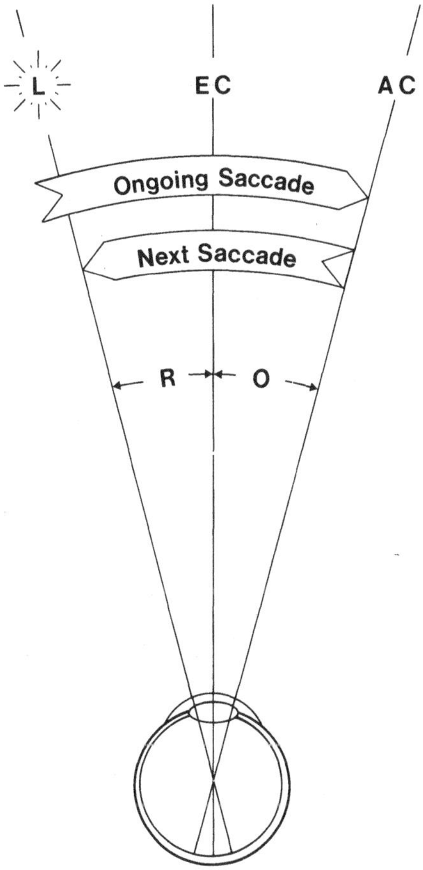

Fig. 2. A point light source, L, is flashed briefly during a clockwise saccade to an intended orientation, AC, represented by an afference copy. The actual eye orientation at the time of the flash, EC, is sensed by means of an efference copy. According to the sum-of-errors hypothesis, the oculomotor control system calculates the size of the next saccade by summing R, the retinal eccentricity of the image, with O, the oculomotor error signal (AC-EC) at the time of the flash. (Figure and quoted caption from Hershberger, 1987, Perception & Psychonomics, 41, 35-44, reprinted with permission of Psychonomic Society, Inc.)

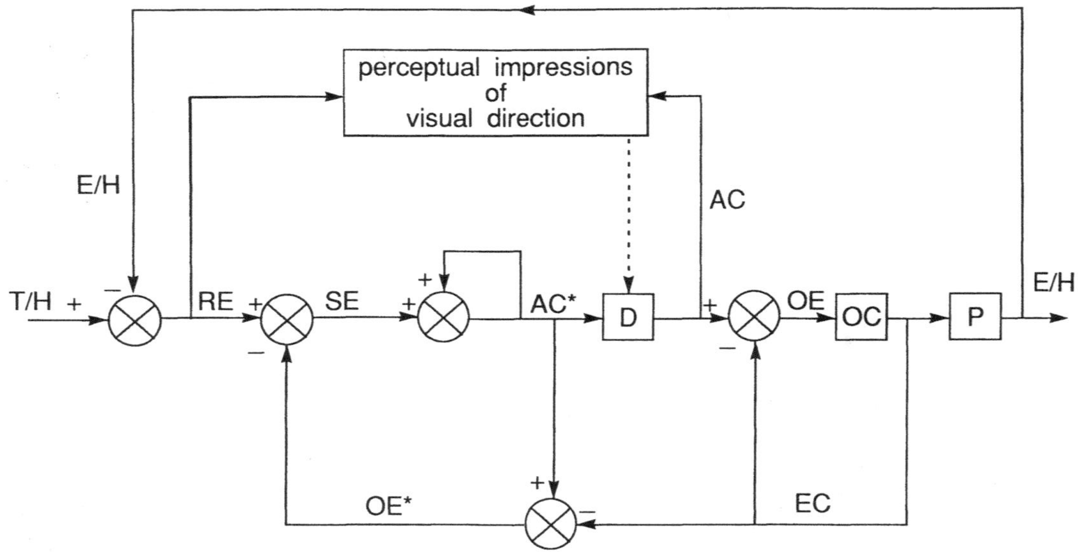

Fig. 3. Modified version of Robinson's egocentric model of the saccadic system. T/H, target position relative to the head; E/H, orientation of the eyes in the head; RE, retinal error (retinal eccentricity of the target's image); D, delay; P, plant; OC, oculomotor components; EC (efference copy), sensed orientation of the eyes in the head; AC (afference copy), neural representation of intended eye orientation relative to the head; AC*, neural estimate of target position expressed as an impending value of AC; OE (oculomotor error), the difference between the intended and sensed orientation of the eyes in the head (AC-EC); OE* (impending oculomotor error), the difference between the to-be-intended and the sensed orientation of the eyes in the head (AC*-EC); SE (sum of errors), the sum of the two types of errors topographically coded by cells in the superior colliculus: +RE, coded by visual cells in the superficial layer, and -OE*, coded by quasi-visual cells in the intermediate layer (Mays & Sparks, 1980a). Perceptual impressions of visual direction are a joint function of RE and AC. The dashed line indicates that a perceptual impression of visual direction, as well as AC*, may serve to specify the impending value of AC.

Suppose that a monkey is gazing 5 degrees to the right of his head's mid-sagittal plane (i.e., AC* = AC = EC = +5 degrees). Further, suppose that a visible target is presented 5 degrees to the left of his head's mid-sagittal plane, 10 degrees to the left of the current value of AC* = AC = EC = +5 degrees. The retinal eccentricity (RE) of this target is -10 degrees whereas the oculomotor error (both OE* and OE) is O. The sum of errors is, hence, -10 degrees, so AC* is decremented by that amount. Now AC* = -5, and EC = +5; so, OE* = -10, and SE = O. Since SE = O, AC* remains fixed at -5 degrees, providing the visible target does not move; and, after the customary saccadic latency, this value of AC* is passed on to the oculomotor loop (to the right of D) to become the new value of AC, and the direction of the monkey's gaze shifts saccadically to the new value of AC: 5 degrees to the left of the head's mid-sagittal plane.

Further, if an electrical stimulus applied to the monkey's superior colliculus during the latent period moves the monkey's eyes, without erasing his immediate memory for this value of AC*, the monkey should still look toward the remembered target, once the effects of the electrical stimulus have passed, just as Mays and Sparks (1980b; 1983b) have found. A shock-elicited saccade of this sort should generate a temporary oculomotor error signal (OE*) in the colliculus (it does: Sparks & Porter, 1983, found that quasi-visual cells respond prior to compensatory movements in their movement fields), but this aberration should not affect an impending saccade to a previously flashed target, because the sum-of-errors mechanism locates *visual* targets and should operate only when a saccadic target is imaged on the retina. For instance, Hallett and Lightstone (1976a) have noted that "the occurrence of a corrective saccade in the dark...is critically dependent upon the target being lit...during at least the early part of the prior primary saccade" (p.101).

In fact it appears that neither AC* nor AC is necessarily altered by electrical stimulation of the superior colliculus. Sparks and Mays (1983b) found that when electrical stimulation was administered shortly (20-30 ms) after an eccentric visual target had flashed, the monkey first looked back to the initial fixation target (now invisible) represented by AC before making a second saccade to the location of the eccentric flash represented by AC*. Therefore, Sparks and Mays (1983a, 1983b) have suggested that electrical stimulation of the superior colliculus directly alters the value of OE. When a monkey is required to look at a remembered target location, as in Mays and Sparks' double-saccade trails, the value of AC* representing the remembered eye orientation is recalled rather that recalculated.

Nonetheless, the oculomotor error signal (OE*) during the saccadic latency would reflect the metrics of the impending saccade to the remembered orientation. That is, a population of neurons coding the value of OE* would be indifferent to the origin of AC*'s value (recalled vs. calculated). In other words, they would resemble the population of quasi-visual cells in the superior colliculus.

Thus, the apparent paradox is resolved. Inasmuch as the reference signal for the oculomotor control loop (i.e., AC) is provided by the target-locating process (i.e., AC*), the target-locating process already "knows" the current value of AC. Hence, given this basic information, all that the superior colliculus need do in order to locate a new saccadic target in head coordinates is register two error signals, and this it clearly does (Mays & Sparks, 1980a, Sparks, 1986; Sparks & Mays, 1983a, 1983b; Sparks & Porter, 1983). The visual cells in the superficial layer topographically code retinal error (RE). The quasi-visual cells in the intermediate layer topographically code oculomotor error. Sparks and Mays (Sparks, 1986; Sparks & Mays, 1983a, 1983b) have suggested that the quasi-visual cells register the value of *OE*, which they call the *impending* oculomotor error. We suggest, instead, that *OE* represents *current* oculomotor error, and that quasi-visual cells code *impending* oculomotor error, *OE**. Although Sparks and Mays have not distinguished between current and impending oculomotor error, (the former defined in terms of AC and the latter in terms of AC*), their findings are perfectly consistent with the distinction, particularly the following finding, also mentioned above (Sparks & Mays, 1983b): When an electrical stimulus is applied to the superior colliculus early during the latency of a saccade to an eccentric flash, the eye first looks back to the initial fixation target (represented by AC) before making a second saccade to the location of the flash (represented by AC*).

We will present evidence below that implies that the value of AC* is typically passed on to the oculomotor control loop about 80 ms before the impending saccade, with the saccade being inhibited for an additional 80 ms by "pause cells" known to be part of the oculomotor loop (Robinsion, 1981). We believe that the "saccade-related" cells in the deep layers of the superior colliculus (Mays & Sparks, 1980a) may also be involved in the process by which AC* values calculated in the target-identification loop are passed on to the oculomotor loop to serve as updated values of AC. Our reasons for this belief are threefold: (a) the "saccade-related" neurons are in the right place: the superior colliculus, (b) they fire at the right time: a discrete burst beginning approximately 20 ms before saccade onset (Sparks, Mays, &

Pollack, 1977), and (c) they topographically code the right signal: the difference between AC and AC* (since the value of this difference will be equivalent to the value of OE*, the quasi-visual cells and the saccade-related cells *appear* to code the same information).

8. Interpreting the Findings of Hallett and Lightstone

The sum-of-errors mechanism also accounts for the findings of Hallett and Lightstone (1976b) who monitored the eye movements of two human subjects attempting to fixate a spot of light that abruptly shifted its position on an oscilloscope screen, the screen itself being invisible. At the beginning of each trial the spot of light remained stationary at the center of the screen directly in front of the left eye (the right eye was occuluded). After a random delay, the spot stepped 7.65 degrees to the left or right to elicit a refixation saccade. Shortly after this refixation saccade began (i.e., 8-12 ms, when the eye had moved only 6-10 minutes of arc), the spot was turned off for 10 ms and then re-presented for 20 ms at a position randomly selected from among four screen positions, ±3.83, and ±11.5 degrees from the screen's center. Although this 20 ms flash appeared only *during* the saccade and illuminated a retinal track of roughly 3-4 degrees in length, the subject refixated the appropriate target location with a subsequent goal-directed saccade of normal amplitude and normal latency. This result would not be expected if retinal image position were the sole information used, but it is entirely consistent with the sum-of-errors hypothesis.

According to the sum-of-errors hypothesis, the process is as follows. At the beginning of the trial when the subject's left eye is in the primary position (O) fixating the spot of light in the center of the screen, the relevant variables assume the following values: RE = O, AC* = AC = EC = O, and OE = OE* = O. When the target steps 7.65 degrees to the right, for example, the value of RE immediately becomes +7.65 degrees. Since OE* = O at this time, the sum of errors is also +7.56 degrees and the current value of AC* is incremented to +7.65 degrees, to become the impending orientation of the eyes. After the customary saccadic latency this value of AC* is passed on to the oculomotor loop (to the right of D) as the new value of AC, and the eye begins its saccade. Now, suppose that the 20 ms flash comes on when the eye has moved 2 degrees to the right (EC = +2) and goes off by the time EC = +6 degrees. Suppose, further, that the 20ms flash emanates from the -3.83 screen location. Initially, when the flash begins, RE = -5.83 degrees, and OE* = 5.65, so SE (i.e, RE - OE*) = -11.48. The feedback loop mulls this sum of errors by decrementing the current value of AC* (+7.65) by this amount, yielding a new value of AC* equal to -3.83, the screen location of the 20 ms flash. As the eye continues to turn, the values of RE and OE* remain equal to each other so SE remains at zero. After the customary saccadic latency the calculated value of AC* is passed on to the oculomotor loop (to the right of D) as the new value of AC, and the eye saccades to the location of the 20 ms flash.

9. Honda's Failure to Replicate Hallett and Lightstone

Honda (1989) has recently reported experimental findings which appear to disagree with those of Hallett and Lightstone (1976a, 1976b). Honda asked his subjects to do two things each trial: (a) to look toward a visual target flashed briefly during a prior saccade (oculomotor localization), and (b) to move a point light source to the location from which the prior flash appeared to come (perceptual localization). Honda found that his subjects tended to move their eyes to the perceptually judged position of the flash, which was illusory, rather than to the actual position of the flash, as had been found by Hallett and Lightstone.

There are two ways in which Honda's subjects could have performed their complex task. Although they were required to move their eyes immediately before they moved the continuous point source to the flash's perceived location, they could have judged the perceptual location of the flash either before or after the eye movement. They could have, first, pointed their eyes reflexively in the direction of the flash and then moved the continuous point source until they perceived it to be oriented in the same direction as their gaze; or, they could have, first, judged the direction of the flash perceptually and then pointed their eyes in that same direction. Whichever type of localization occurred first (oculomotor or perceptual) would evince the lesser variability. Since the variance of the oculomotor localizations was greater, it appears that Honda's subjects were doing the latter. That is, they appear to have been pointing their eyes in the direction from which they had perceived the flash to have come. This being the case, one would expect their oculomotor localizations, as well as their perceptual localizations, to evince the perceptual errors that are known to occur when subjects make psychophysical judgments of the visual direction of perisaccadic light flashes (e.g., Matin 1972). In general, they did; the intrasaccadic flashes which occurred early in the eye movement appeared to be displaced in the direction of the eye movement--although flashes which occurred late in the eye movement appeared displaced in the opposite direction.

Consequently, Honda's failure to replicate Hallett and Lightstone's findings is more apparent than real. The different results are attributable to the two different methods employed. Hallett and Lightstone's subjects were simply attempting to fixate a single spot of light; they were uncertain as to whether the target remained continuously lit and made poor guesses about their eye movements. Their eye movements appear to have been driven reflexively by the disturbances to the light spot's position on the oscilloscope screen. In contrast, Honda's subjects appeared to have been judging the perceived visual direction of an intrasaccadic flash, and pointing their eyes in that same direction. Incidentally, although the saccade during which the judged flash occurred was itself triggered by a brief luminous target in Honda's experiment, that prior target, a pair of rectangular LEDs, differed markedly from the judged intrasaccadic flash, which came from a round LED. In effect, Honda's subjects, like Matin's, were making psychophysical judgments of visual direction. Honda's experiment used Fechner's method of adjustment, whereas Matin's experiments used Fechner's method of constant stimuli (for an explanation of Fechner's psychophysical methods, see Woodworth & Schlosberg, 1954). But inasmuch as these two psychophysical methods are merely alternate modes of measurement, it is not surprising that the experimental results of Honda and Matin are very similar. Thus, the findings of Honda and Matin, on the one hand, and Hallett and Lightstone, on the other, remain paradoxical. In the next section we will explain how the afference-copy hypothesis resolves this paradox.

10. Matin versus Hallett and Lightstone: A Resolution

We suggest that the most reasonable resolution of the paradoxical findings of Matin and his associates on the one hand and Hallett and Lightstone on the other is to be found in the familiar distinction between the retinotectal and the geniculostriate divisions of the visual system (Schneider, 1969; Trevarthen, 1968). We submit that the two systems estimate E/H differently. As an integral part of the sum-of-errors hypothesis detailed above, the retinotectal (i.e., collicular) system appears to employ EC (Guthrie, Porter, & Sparks 1983) as an estimate of E/H (i.e., OE* = AC* - EC). As hypothesized below, the geniculostriate system mediating perceptual impressions of visual direction appears to employ AC as an estimate of E/H. Hershberger (1987) has called this the afference-copy hypothesis (AC hypothesis).

The Phenomena Requiring Explanation

Matin (e.g., Matin, 1972, 1982; Matin & Pearce, 1965) has found that the perceived visual direction of a perisaccadic flash coming from a point light source viewed in the dark appears displaced in the direction of the attendant saccade, and that this illusory displacement occurs even before the eyes begin to move. The clue to understanding this illusory visual direction is provided by the fact that it appears even before the eyes begin to move. It is as if perceived visual direction depends upon the intended direction of gaze (represented by AC) rather than upon the actual direction of gaze (estimated by EC). That is, the shift in retinal local signs (i.e., the spatial coordinates of the retinal input) precedes the eye movement, just as the intention to move the eyes precedes the eye movement. This interpretation (i.e., the AC hypothesis) is illustrated in Figure 3.

The Afference-Copy Hypothesis

According to the AC hypothesis, the spatial information mediated by the retina (expressed in retinotopic coordinates) is combined with information about the orientation of the eyes (expressed in head coordinates) in order to mediate conscious visual impressions of the environment, expressed in the egocentric coordinates of the head (and, because of the vestibular ocular reflex, the allocentric coordinates of the earth, as well). According to the AC hypothesis, the extraretinal signal representing eye orientation is the current value of AC, not the current value of EC.

The AC hypothesis may be viewed as a particular version of Helmholtz's (1867/1962) "effort of will". However, it differs from the received view of Helmholtz's theory, which is von Holst and Mittelstaedt's (1950) efference copy hypothesis or Sperry's (1950) notion of corollary discharges of efference. According to the AC hypothesis, the spatial coordinates of the retinal input (i.e., local signs) are shifted discretely, because AC is updated in discrete shifts. According to the received view of Helmhotz's theory, the visual frame of reference should undergo a continuous shift which is as gradual as a saccadic eye movement.

The magnitude of the perisaccadic illusions of visual direction, as defined by Matin's data, is time dependent (e.g., Matin, 1972, Figure 9); that is, the magnitude varies as a continuous function of time, beginning prior to the saccade itself. Frequently, Matin's data have been interpreted as meaning that perisaccadic shifts in retinal local signs also occur gradually as a continuous function of time. However, there are both retinal and empirical reasons to believe that that interpretation is incorrect. The various empirical reasons will be presented throughout the remainder of the chapter. The rational reason, entailing a methodological consideration, is summarized in the next section.

The Limitations of the Method of Constant Stimuli

The psychophysical procedure that Matin and his associates have employed in their many experiments is known as Fechner's method of constant stimuli (Woodworth & Schlosberg, 1954). Specifically, Matin's subjects were required to say whether a perisaccadic test flash presented on each trial appeared to be located to the left or to the right of a prior fixation target (saccades were horizontal). This psychophysical method is insensitive to discrete shifts of retinal local signs, and will even misrepresent a discrete shift as a continuous one:

> For instance, suppose, for sake of argument, that the shift in retinal local signs that attends a saccade occurs in a discrete, stepwise fashion, and that the latency, but not size, of this step varies from trial to trial. Successive trials of repeated stimulation of the same retinal locus at the same relative time (e.g., 20 msec prior to eye movement) will yield a bimodal distribution of apparent visual direction, one mode comprising the effects of the trials on which the stimulus precedes the shift and the other comprising the effects of the trials on which the stimulus follows the shift. The central tendency of the distribution as a whole, customarily taken to represent the true visual direction or local sign of the retinal signal, may be observed to depend heavily upon the relative frequencies of the two types of trials, which, in turn, depend heavily upon when during the perisaccadic interval the stimulus is presented. In general, the later the stimulus occurs in the interval, the more frequent the postshift trials are likely to be and, hence, the greater the apparent shift in local signs, even though the actual shifts are all of the same magnitude whenever they occur (Hershberger, 1987, p. 40).

Since Honda(1989) used Fechner's method of adjustment, the temporal aspects of his data may be taken at face value; and, when we look at the time course of the illusion of visual direction defined by Honda's psychophysical data, we find that it is very brief: 30ms. But this brevity is also a methodological accident: Honda's test flashes were all presented *during* a prior saccade, so the observed time course could not possibly exceed the duration of the prior saccade. All this is to say that neither Matin nor Honda have determined the exact time course of the shift in retinal local signs that mediates the perisaccadic illusion of visual direction.

11. Empirical Evidence for the AC Hypothesis

There is a variety of extant data that implies that retinal local signs shift discretely, before the saccade. A relatively recent experiment by O'Regan (1984) is very suggestive. O'Regan used a spot of light on a cathode ray tube to stimulate the fovea of his subject's retina at various times either before, during, or after a saccadic eye movement. The subject then moved a screen cursor to the perceived position of the flash. One of O'Regan's 3 subjects always located the foveal flash either at the fovea's departure point or at its arrival point and never near its veridical position in between. This, of course, is precisely consistent with the AC hypothesis. O'Regan's second subject performed similarly, although his settings showed more variability. The data from the third subject were very noisy and difficult to interpret.

Because O'Regan's initial and terminal fixation targets (small luminous triangles) remained visible throughout each trial, the retinal image of the terminal flash eventually became superimposed upon the retinal remanence of the test flash (the letter I) imaged on the fovea during the saccade. Therefore, seeing the two (the I and the triangle) as occupying the same position in space may merely have reflected the fact that they both shared the same retinal locus at trials end. Thus, although his data are consistent with the AC hypothesis, they do not provide definitive evidence that retinal local signs change discretely as the intended direction of gaze changes from one orientation directly to another. However, there is such evidence, some of it being of a traditional variety, involving judgements made by subjects whose extraocular muscles have been paralyzed, and some of it being of a new variety involving the phantom array (Hershberger, 1987). We will consider the former variety first.

Paralysis of the Extraocular Muscles

Helmholtz argued that the perceived visual direction of visible objects depends upon the *intended* direction of gaze because an object whose image is stabilized on the retina (e.g., a negative afterimage) appears to be located in whatever direction one intends to gaze, even when the extraocular muscles are paralyzed. He cited von Graefe (1854), an opthalmologist who had examined a patient with an acute palsy of the lateral rectus of one eye.

When this patient attempted to look in a direction requiring a contraction of this paralysed muscle, stationary objects imaged on the retina of the effected eye (the mobile eye being occluded) shifted their apparent visual direction in the direction of the intended eye movement. Hence, the patient would reach beyond these objects when attempting to grasp them (past pointing).

> Both of these effects of *partial* extraocular paralysis (the illusory visual shift and the past pointing) have been observed repeatedly since that time, both by other clinicians (Cogan, 1956; Jackson & Paton, 1909) and in controlled experiments (Brindley & Merton, 1960; Kornmueller, 1931; Siebeck, 1953, 1954; West, 1932). These effects are consistent with both versions of Helmholtz's theory, the efference copy hypothesis and the AC hypothesis.
>
> As a test of the AC hypothesis, the most telling findings are those of Stevens et al. (1976) who recently examined the effects of *total* extraocular paralysis under controlled experimental conditions. According to the AC hypothesis, they should have found that each intended change of gaze should have altered the subject's local retinal signs in a discrete, stepwise fashion.
>
> The scene need not appear to *move* as such, at least not any more than it does when the eye moves normally, which is to say not at all. It need only appear to be egocentrically *displaced* in the sense that any thought of reaching out and touching visible objects lying on the line of sight would seem to require an arm extension in a direction consistent with the intended direction of gaze, and that if the person were indeed able to extend his or her arm in this direction his or her reaching out would exhibit "past pointing." Of course, past pointing at eccentric visual targets is not an inevitable consequence of extraocular paralysis. Just as it is possible to point one's hand and arm accurately at a retinally eccentric target viewed with an immobile normal eye staring straight ahead, so it would be possible to do so with an immobile paralyzed eye staring straight ahead. Past pointing is to be expected only to the degree to which the direction in which one is pointing the hand is specified by erroneous *extraretinal* information (Hershberger, 1987, p. 42).

Stevens et al. administered two different pharmacological agents to one subject (JKS) in separate experimental sessions. In some sessions curare/succinylcholine, a neuromuscular blocking agent, was administered systemically. In other sessions, procaine was injected directly into the extraocular muscle cone. In both cases the experimental findings were consistent in detail with the above predictions. After the first systemic injection of succinylcholine,

> JKS reported no *movement* or *displacement* during attempted saccades. "I tried to move my eyes as hard as I possibly could and nothing happened, the world was just there....I simply could not move my eyes." [After both a second and a third systematic injection, JKS again]... reported that he was very much aware that his eyes were paralyzed. "I know I did not move my eyes. I was trying very hard." However, unlike the first total paralysis experiment, "When I looked to the right I felt that if I had to touch anything...I would have to reach over to the right" (Stevens, et al., 1976, p. 95)

JKS felt that this illusory displacement of the scene was much the same as that observed during an earlier experiment in which he had been only partially paralyzed, except that with partial paralysis the displacement was accompanied by a "rapid jerk" or "jumping" of the scene.

> When total extraocular paralysis was achieved [by means of retrobulbar injection of procaine] JKS reported the same perception of *displacement* without noticeable jumping, as seen in the succinylcholine experiments....Past pointing during the total block was very strong. During one study JKS attempted to touch an object in the periphery and overshot by 20 in. (Stevens et al., 1976, p. 96)

Although these findings of Stevens et al. are precisely consistent with the AC hypothesis, they are inconsistent with the findings of Siebeck (1953, 1954; Siebeck & Frey, 1953) and Brindley, Goodwin, Kulikowski, and Leighton (1976), who failed to find any displacement or past pointing effects resulting from total extraocular paralysis. It is possible that the subjects in these two experiments merely failed to note the displacement, just as JKS had failed to do in his first experimental session; and past pointing is not an inevitable consequence of paralysis. Also, Matin et al. (1982) have shown that a well-illuminated visual field serves to mask some of the perceptual effects of partial paralysis and have suggested that it may similarly influence the effects of full paralysis as well. Therefore, Matin et al. have recommended that the total-paralysis experiments be replicated in darkness, implying that the effects reported by JKS would be fully replicated. To the degree that their expectations are warranted, the total-paralysis experiments provide compelling evidence for the AC hypothesis, which alone predicts displacement without jumping.

The Phantom Array

Perhaps the most compelling evidence for the AC hypothesis is the phantom array, a complex perisaccadic illusion of visual direction first described by Hershberger (1987).

> If one fixates alternately to the left and right of a rapidly blinking light viewed in the dark, one will see a spatially extended series of phantom lights blinking on and off sequentially in a direction opposite to that of the saccade. The direction of this motion appears merely to reflect the direction of the retina's motion across the blinking image. What is remarkable is that the perceived shift in the direction of regard is reflected only in what appears to be a discrete *displacement* of the entire array of phantom lights in the direction of the eye movement. For instance, if the arrow in Figure [4] represents a single saccadic eye movement and the asterisk represents a single flashing light, the bracketed array represents the phenomenal appearance. One sees, fixed in space, a horizontal array of lights that blink on and off in sequence, giving an impression of apparent motion, or *phi*. The phantom array does not itself appear to move; however, neither is it centered upon the light. Rather, in the case illustrated above, the array is displaced to the right, with its left end appearing to be located in the light's presaccadic direction. Because the flash seen on the right end of the phantom array is painted onto the retina first, that is, before any of the other flashes seen in the array and because the gaze continues to shift to the right as the remaining flashes in the array are being painted onto the retina, the retinal locus of the remanence of the first flash moves through a substantial visual angle equally as large as the phantom array itself. If the sign (perceived visual direction of that retinal locus) shifted continuously and isometrically with the eye movement, then the first flash should appear to move rightward in the direction of the changing gaze by an angular amount equal to that subtended by the phantom array. But it does not appear to move to the right at all. Rather, it appears to be displaced, or placed to the right by the observed amount all at once without having moved through the intermediate locations. Neither are any of the other flashes in the array seen to move to the right. They are seen to be placed to the right but not to be moving to the right. This is true in spite of the fact that the

saccade does not itself preclude motion perception: phi to the left is clearly visible within the phantom array. This implies that changes of perceived direction of regard are either entirely presaccadic or very abrupt--or both (pp.38-39).

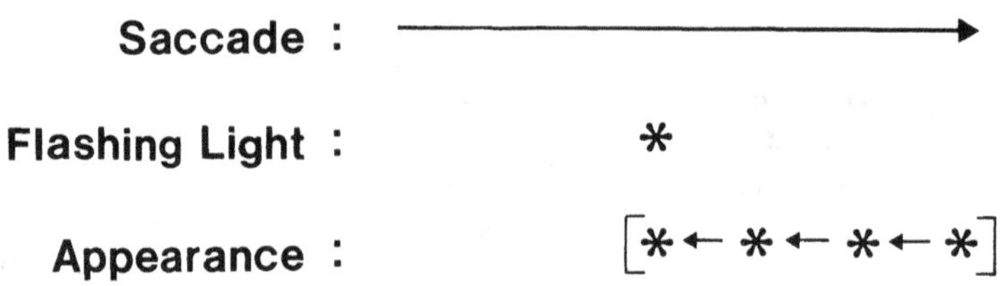

Fig. 4. If you shift your gaze saccadically from the left to the right of a point light source in a darkened room, blinking on and off at 120 Hz, you will see phi movement to the left within a phantom array that is displaced to the right. (Figure and quoted caption from Hershberger, 1987, Perception & Psychonomics, 41, 35-44, reprinted with permission of Psychonomic Society, Inc.)

Although Hershberger (1987) described in precise detail both the complexity and the profound theoretical implications of the phantom array, he did not demonstrate the replicability of the phenomenon across observers. Obviously, the theoretical significance of the phenomenon will depend upon its replicability. To determine whether the phantom array is replicable across observers, we (Jordan & Hershberger, in preparation) have individually tested 75 naive observers. The observers were carefully interrogated to determine whether they experienced the complex phantom array in all of its aspects. As described above, the phantom array has the following four aspects:

1. For horizontal saccades, the phantom array comprises a horizontal row of dots. We will refer to this aspect as ROW.

2. The dots within the array materialize in a regular sequence in the direction opposite the eye movement. We will refer to this aspect as SEQUENCE.

3. The array of dots does not appear to move in the direction of the eye movement. We will refer to this aspect as FIXED.

4. The entire array of dots appears to be located to one side of the blinking light, the side associated with the new direction of gaze. We will refer to this aspect as SIDE.

The research is summarized in the following section.

12. Testing the Phantom Array

General Method

A total of 75 undergraduates enrolled in Introductory Psychology at Northern Illinois University participated as naive observers. In order to obtain their informed consent, they had been told that their task would be to describe the appearance of some lights viewed in the dark. Further, they were told that they would receive bonus course credit for participating whether or not they completed the task.

The observers participated one at a time. They sat at a table, with their arms resting on the table and their heads supported by a chin rest. A red, light-emitting diode (LED) oscillating on and off at 200 Hz (cycle: 1 ms on/4 ms off) was located 160 cm in front of the chin rest. At this distance, the 5 mm (diameter) LED subtended a visual angle of 1.8 degrees; and, oscillating at 200 Hz, it did not appear to flicker. The room was totally dark except for the LED. From a adjacent room, the experimenter communicated with the observer via an intercom.

To begin the interrogation, the experimenter asked the observers whether they could see the red light. When they replied in the affirmative, the experimenter read the first set of instructions (Instruction #1) printed below. (Since the value of our observer's answer are only as trustworthy as our method of interrogation, it is important to describe our method of interrogation in detail.)

> *Instruction #1:* I want you to quickly look at the left corner of the room then the right corner of the room, then back to the left corner and so on. Once you get your eyes to either corner, keep them there for a second before looking to the other corner. I know you can't see the corners of the room but pretend that you can. While you are looking back and forth between the corners of the room, make sure that you do both of the following; (1) keep your head still, and (2) move your eye *across* the red light. Don't look at the red light, just make sure your eyes move across it. While doing this task, note what happens to the red light. Continue looking from corner to corner until you have a good idea of what you are seeing. When you have a good idea of what you are seeing, comment on it. It will then ask you a few questions.

Do Naive Observers See ROW?

After the observers indicated that they had a good idea of what they were seeing, the experimenter asked them five forced-choice questions which had been designed to determine whether or not the observer was seeing, during each eye movement, a horizontal row of dots. (ROW). Each question in this series of 5 could be asked only if the questions that preceded it has been answered in a manner congruent with seeing a horizontal row of dots during the saccade. Thus, if for any of these first five questions the observer selected the answer that was non-congruent with seeing a horizontal row of dots, the experimenter repeated Instruction #1 and asked the observer the question again. If, after one repetition, the observer still selected the non-congruent response, this response was recorded as a final datum and the observer was excused. The five questions and the observer's *initial* answers were as follows:

Q#1. Each time you move your eyes, do you see any dots or flashes? Results: 74 (99%) initially responded "yes."

Q#2. Each time you move your eyes do you see only one dot or more than one dot? Results: 73 (97%) initially responded, "more than one."

Q#3. Do all the dots appear at one point in space or do they appear to be spread out? Results: 72 (96%) initially responded, "spread out."

Q#4. Is the spatial arrangement of the dots that you see random or regular; that is, do you see a shapeless cluster of dots, or a regular pattern such as a dotted line? Results: 72 (96%) initially responded "a dotted line."

Q#5. Is the dotted line vertical as in up and down or horizontal as in side to side? Results: 72 (96%) initially responded, "horizontal."

One observer never reported being able to see more that one dot during any given eye movement (i.e., twice he answered question Q#2 in a non-congruent manner). Another observer, who had initially answered each question (Q#1-Q#5) in a manner congruent with seeing a horizontal row of dots, insisted that she was no longer able to see the phenomenon when it came time to answer question Q#6. Both of these observers were dismissed. Thus, the number of observers going on to question Q#6 was 73.

Do Naive Observers See SEQUENCE?

Once the experimenter was sure that an observer was reliably seeing a horizontal row of dots during every saccade, he then determined whether or not the dots in the row appeared to materialize in a regular sequence opposite the direction of the saccade (SEQUENCE). The experimenter accomplished this by reading Instruction #2 to the observer, followed by three more forced-choice questions, Q#6, Q#7, and Q#8.

> *Instruction #2:* Now I would like for you to look as far to the right as possible then as far to the left as possible while keeping your head still. Continue to do this at a steady pace until you can answer the following question. (Insert the appropriate question)

Instruction #2 was designed to accentuate the phenomena being investigated by these three questions. If, for question Q#6, Q#7, or Q#8, an observer selected the response that was non-congruent with seeing the horizontal row materialize in a regular sequence opposite the direction of the saccade, the experimenter noted this, skipped the rest of the questions in this section, and proceeded to questions Q#9-Q#11, which were not dependent on the observer's answers to questions Q#6 through Q#8. However, after asking those questions (Q#9-Q#11), the experimenter reread Instruction #2, and returned to the particular question to which the observer had initially given a non-congruent answer (Q#6, Q#7, or Q#8) a second time. This circuitous procedure was followed in order to minimize the probability of the observer simply switching his/her response to satisfy the experimenter, as might be expected if a question had been repeated immediately following a non-congruent answer. If after returning to this section the observer still gave the non-congruent response, the experimenter recorded this and marked the remaining questions of this section as not answered (NA).

Q#6. Each time you move your eyes, do some of the dots appear before others, or do they all appear at once? Results: 45 of the 73 observers (62%) initially responded, "some appear before others." Of the 28 who initially responded, "all at once," 15 (54%) changed their mind when asked to observe the phenomenon a second time.

Q#7. Is the sequence in which they appear random, meaning without any type of order, or is the sequence regular as in left to right or right to left? Results: Of the 60 observers asked this question, 56 (93%) initially responded, "in a regular sequence." Of the 4 who initially responded, "in a random sequence," 2 changed their mind when asked to observe the phenomenon a second time.

Q#8. You stated that the dots appear in a regular sequence. Is this sequence in the same direction or the opposite direction of your eye movement? Results: Of the 58 observers asked this question 45 (78%) initially responded, "in the opposite direction." Of the 13 who initially responded, "in the same direction," 8 (62%) changed their mind when asked to observe the phenomenon a second time.

Do Naive Observers See FIXED?

Question Q#9 was designed to determine whether or not the position of a single dot in the array remained fixed while the array was present (FIXED).

Since this question was only concerned with the behavior of a single dot within the array, it could be asked regardless of how the observer had answered questions Q#6 through Q#8. The experimenter reread Instruction #2, and then asked the following question:

Q#9. You stated that each time you move your eyes you see a horizontal row of dots. Once an individual dot in the row materializes, does its location change, or does it appear fixed at that location? Results: Of the 73 observers asked this question, 28 (38%) responded, "remains fixed in space." The 40 observers who responded with "changes location" were asked the following follow-up question:

Q#9a. Does the dot's location change in the same or the opposite direction of your eye movement? The experimenter asked this question because it was assumed that some observers would confuse the phi produced by the sequential materialization of the dots, with the location of a single dot. Results: Of the 45 observers asked this question, 40 (89%) responded, "in the opposite direction."

Combining the results of these two questions (Q#9 & Q#9a), 68 (93%) of the 73 observers saw the dots either as being stationary or as moving in the direction opposite the eye movement.

Do Naive Observers See SIDE?

Instruction #3 and questions Q#10 and Q#11 were designed to determine whether or not the entire array appeared on the side of the array associated with the new direction of gaze (SIDE).

> *Instruction #3*: Now I would like you to do what I call the LEFT-TO-RIGHT TASK. First, look at the left corner of the room. Than look rapidly to the right corner of the room. After you have done this, close your eyes and move them back to the left corner of the room. Once you are there, open your eyes and repeat the entire process. Continue doing this until you can answer the following question: (Insert the appropriate question).

The experimenter had observers perform the left-to-right eye movement in order to simplify the communication process. If the observers had kept their eyes open while producing leftward and rightward saccades they would have had to inform the experimenter about the direction of their eye movement before commenting about the location of the array.

Q#10. Each time you move your eyes, does the row of dots appear on both sides of the red light or just one side of the red light? Results: Of the 73 observers asked this question, 57 (78%) initially responded, "one side." Of the 16 who initially responded, "both sides," 10 (63%) changed their mind when asked to observe the phenomenon a second time.

Question Q#11 could not be asked unless the response to question Q#10 had been "one side". If the observer answered Q#10 with "both sides," the experimenter repeated Instructions #3 and repeated question Q#10; however, first the experimenter returned to any previous questions left unanswered (Q#6, Q#7, or Q#8) starting with a repetition of the one to which the observer had initially given a non-congruent answer. If when question Q#10 was repeated the observer answered "both sides", the experimenter recorded this and marked question Q#11 as NA. If the observer answered question Q#10 with "one side", the experimenter recorded this and asked him/her question Q#11.

Q#11. You said that all the dots appear on one side of the light. Which side, the right or the left? Results: Of the 67 observers asked this question, 49 (73%) initially responded, "the right side." Of the 18 observers who initially responded, "the left side, 8 (44%) changed their mind when asked a second time.

If the observer responded with "the left side" to question Q#11 the experimenter returned to Instruction #2 and repeated any questions that required repetition (i.e., Q#6, Q#7, and Q#8) before repeating Instruction #3 and repeating question Q#11.

Summary of the Results

All but 2 of the 75 naive observers consistently saw a horizontal row of flashing dots whenever they saccaded horizontally across the flashing LED, and even one of these 2 people initially appeared to see the array. Further, of the 60 observers who were able to see the temporal sequence of the flashes, all but 4 saw them flashing on and off sequentially in a direction opposite to the eye movement's, and even 2 of these 4 observers changed their mind they were asked to examine the phenomenon a second time (i.e., they too reported seeing a direction of motion opposite to the eye movement's). Furthermore, of the 73 observers who consistently saw the horizontal row of dots, all but 5 (i.e., 93%) saw the individual dots as being stationary or as moving in a direction opposite to the eye movement's (this apparent motion appears to be the phi produced by the sequential flashes).

(The 5 observers who reported that the dots moved in the same direction as the eye movement may have been reporting a displacement of the entire array rather than a motion of the individual dots, because all 5 of these observers subsequently reported that they saw array displaced to the right when they looked to the right.) Finally, only 6 (8%) of 73 observers consistently saw the array as straddling the location of the LED. The other 67 observers saw it located to one side of the LED, with only 10 (15%) of these 67 observers consistently locating it on the side opposite the new direction of gaze. That means that 57 (85%) of these 67 observers saw the entire array on the side associated with the new direction of gaze.

As a rule, the vast majority of our naive observers reported seeing the phantom array essentially as Hershberger (1987) described it. Irrespective of whether the minority of incongruent answers represent fundamental individual differences or merely experimental noise, it is clear that the phantom array is a robust perceptual phenomenon that is replicable across most observers.

The fact that the phantom array observed in this experiment was generated by a 1 ms flash occurring once every 5 ms implies that the shift in retinal local signs occurs very rapidly, that is, within 5 ms. Further, since we ourselves continue to see the phantom array when the LED flashes at 500 Hz (i.e., when a 1 ms flash occurs every 2 ms) the shift in retinal local signs is virtually discrete, as the AC hypothesis predicts.

When Does the Shift in Retinal Local Signs Occur

The AC hypothesis implies that the shift may occur before the eye begins to move. This, however, is not a necessary implication of the AC hypothesis. It depends upon the speed with which the geniculostriate system processes the extratretinal information (i.e., registers the new value of AC) mediating the shift in retinal local signs. For instance, the few (6) naive observers in our experiment who reported seeing the phantom array extending on *both* sides of the LED may have experienced a shift in retinal local signs during their saccades (of course, these 6 reports might also reflect measurement error). That is, it is an empirical question as to exactly when the perisaccadic shift in retinal local signs occurs. We have ourselves addressed this question in a recently completed experiment (Jordan & Hershberger, in preparation).

In order to determine exactly when the shift of retinal local signs occurs, we enlisted the assistance of 4 trained psychophysical observers. On each of 1600 trials each observer saccaded from a fixation LED (to the left) to a target LED (to the right) across a blinking LED (200 Hz) which painted the phantom array across the observer's retina. The observer's eye movements were tracked by an infra-red eye-movement monitor. The blinking LED was turned on 5 ms after the saccade began and turned off 1 ms after the saccade ended. The blinking LED was vertically bracketed by a pair of LEDs (VB) that flashed on briefly (1 ms) each trial. The perisaccadic moment when VB flashed was systematically varied in an unpredictable manner across trials. The observer's task each trial was to judge the position of VB relative to the phantom array. The observer used a 5-point scale, with 1 meaning far left and 5 meaning far right. Whenever VB was presented immediately after the observer's retinal local signs has shifted, VB should have appeared to be located at the far right end of the phantom array (position 5). Thus, we were able to use the observer's judgments of the relative position of VB to measure the latency of the extraretinal shift of retinal local signs.

For each of our 4 observers, the extraretinal shift of retinal local signs was observed to occur well before the saccade (about 80 ms, or more), with the latency of the shift varying with the latency of the saccade (we are currently working to refine this measurement). According to the AC hypothesis, this minimal shift-saccade asynchrony of 80 ms implies that the value of AC in the oculomotor loop is typically updated (so as to equal AC*) at least 80 ms before the saccade. That is, the value of AC* in the target-identification loop is passed on to the oculomotor loop as a new value AC typically no later than 80 ms before the onset of the saccade. Immediately, as AC is updated, the retinal local signs change, whereas the saccadic eye movement is inhibited for an additional 80 ms by "pause cells" known to be part of the oculomotor loop (Robinson, 1981).

Becker and Jurgens (1975) have reported a finding which appears related to our finding of an 80 ms shift-saccade asynchrony. They found that the amplitude of a saccade may be altered by retinal information presented as late as 80 ms prior to the saccade. This coincidence suggests that retinal information available during the saccadic latency is utilized until the new value of AC* is passed on to the oculomotor control loop, thereby shifting the retinal local signs. This also implies that retinal information available during the last 80 ms of the impending saccade's latency could affect subsequent "corrective saccades" (Becker & Jurgens, 1975).

Although our observer's perceptual shifts occurred prior to their saccades, their judgments did not always seem to reflect a discrete shift. When VB was flashed during the saccadic latency, the observers sometimes reported seeing VB located at one of the three intermediate positions (2, 3 or 4) and not merely at 1 or 5, as a discrete shift would imply. Further, when an intermediate position was reported, the numerical value of the reported location tended to increase as VB flashed later and later during the latent period, with the modal moment for "3" judgments being about 80 ms prior to the attendant saccade. The latter fact probably reflects the variability of the *shift's* latency, but the former fact seems to imply that retinal local signs shift as a continuous function of time. However, such being the case, the shift appears to be *very* abrupt (i.e., take less than 2 ms) because a LED continuously flashing at 500 Hz (on 1 ms, off 1 ms, etc.) generates a standard phantom array when we (authors) saccade across it.

13. Conclusions and Implications

Taken together, the AC hypothesis and the sum of errors hypothesis account parsimoniously for all three of the aforementioned paradoxes.

The AC hypothesis also accounts parsimoniously for the phantom array, a complex illusion of visual direction that is replicable across most if not all observers. According to the AC hypothesis, retinal local signs shift when the intended direction of gaze (AC) shifts, rather than when corollary efference (ED) is discharged. Because the perceptual shift in retinal local signs precedes the saccade, the shift *introduces* an error of apparent visual direction which the saccade eventually cancels (i.e., the *first* flash in the phantom array appears displaced the *most*). although this may be classified as a version of cancellation or compensation theory, it is the opposite of the received view, which is that a neural signal (i.e., corollary discharge of efference, or efference copy) cancels or compensates for the effects of the eye movement (e.g., Wallach, 1987; however, cf. MacKay, 1973).

Implicit in our sum-of-errors explanation of Hallett and Lightstone's findings is the notion that the colliculus is capable of operating independently of the neural processes mediating conscious perceptual impressions of visual direction. An implication of our explanation is oculomotor "blindsight". For instance, patients with lesions in the striate cortex should be able to respond *reflexively* to luminous targets located within the attendant scotoma even though the patients are perceptually unaware of the targets. Such is the case, in fact. Rafal, Smith, Krantz, Cohen, and Brennan (1990) have recently found that "distractor signals in the blind half of the visual field [of such patients] can inhibit saccades toward targets in the intact visual field" (p. 118). Conversely, voluntary saccades can also inhibit involuntary saccades (Judge, 1973).

References

Baichenko, P. I., Matyushkin, D. P., & Suvorov, V. V. (1968). Participation of fast and tonic oculomotor systems in stretch reflexes and labyrinthine reflexes of extraocular muscles. *Neuroscience Translations, 3,* 350-358. (Translation reprinted from *Fiziologicheskii Zhurnal SSSR imeni I. M. Sechenova,* 1967, *53,* (82-90.)

Becker, W., & Jurgens, R. (1975). Saccadic reactions to double-step stimuli: Evidence for model feedback and continuous information uptake. In G. Lennerstrand & P. Bach-y-Rita (Eds.). *Basic mechanisms of ocular motility and their clinical implications* (pp. 519-524). New York: Pergamon Press.

Boeder, P. (1961). The co-operation of extraocular muscles. *American Journal of Ophthalmology, 51,* 469-481.

Bridgeman B., & Delgado, D. (1984). Sensory effects of eyepress are due to efference. *Perception & Psychophysics, 36,* 482-484.

Brindley, G., Goodwin, G., Kulikowski, J., & Leighton, D. (1976). Stability of vision with a paralyzed eye. *Journal of Physiology* (London), *258,* 65P-66P.

Brindley, G. S., & Merton, P. A. (1960). The absence of position sense in the human eye. *Journal of Physiology* (London), *153,* 127-130.

Bruce, C. J., & Goldberg, M. E. (1981). Frontal eye fields in the monkey: Classification of neurons discharging before saccades. *Neuroscience Abstracts, 7,* 131.

Cogan, D. G. (1956). *Neurology of the ocular muscles.* Springfield, IL: Thomas.

Collins, C. C. (1975). The human oculomotor control system. In G. Lennerstrand & P. Bach-y-Rita (Eds.). *Basic mechanisms of ocular motility and their clinical implications* (pp. 145-180). New York: Pergamon Press.

Cooper, S., & Daniel, P. D. (1949). Muscle spindles in human extrinsic eye muscles. *Brain, 72,* 1-24.

Crommelinck, M., Guitton, D., & Roucoux, A. (1977). Retinotopic versus spatial coding of saccades: Clues obtained by stimulating deep layers of cat's superior colliculus. In R. Baker & A. Berthoz (Eds.), *Control of gaze by brain stem neurons* (pp. 425-436). Amsterdam: Elsevier/North-Holland.

Dichgans, J. (1977). Optokinetic nystagmus as dependent on the retinal periphery via the vestibular nucleus. In R. Baker & A. Berthoz (Eds.), *Control of gaze by brain stem neurons* (pp. 261-268). Amsterdam: Elsevier/North-Holland.

Enright, J. T. (1984). Saccadic anomalies: Vergence induces large departures from ball-and-socket behavior. *Vision Research, 24,* 301-308.

Epstein, W. (1973). The process of 'taking-into-account' in visual perception. *Perception, 2,* 267-285.

Fuchs, A. F., & Luschei, E. S. (1970). Firing patterns of abducens neurons of alert monkeys in relationship to horizontal eye movement. *Journal of Neurophysiology, 33,* 382-392.

Gallistel, C. R. (1980). *The organization of action.* New York: Erlbaum.

Gibson, J. J. (1966). *The senses considered as perceptual systems.* Boston: Houghton Mifflin.

Gibson, J. J. (1979). *The ecological approach to visual perception.* Boston: Houghton Mifflin.

Goldberg, M. E. (1983). Iconoclasm avoided: What the single neuron tells the psychologist about the icon. *Behavioral & Brain Sciences, 6,* 20-21.

Goldberg, M. E., & Bruce, C. J. (1981). Frontal eye fields in the monkey: Eye movements remap the effective coordinates of visual stimuli. *Neuroscience Abstracts, 7,* 131.

Grusser, O. J. (1986). Interaction of efferent and afferent signals in visual perception. A history of ideas and experimental paradigms. *Acta Psychologica, 63,* 3-21.

Guthrie, B. L., Porter, J. D., & Sparks, D. L. (1983). Corollary discharge provides accurate eye position information to the oculomotor system. *Science, 221,* 1193-1195.

Hallett, P. E., & Lightstone, A. D. (1976a). Saccadic eye movements towards stimuli triggered by prior saccades. *Vision Research, 16,* 99-106.

Hallett, P. E., & Lightstone, A. D. (1976b). Saccadic eye movements to flashed targets. *Vision Research, 16,* 107-114.

Hansen, R. M., & Skavenski, A. A. (1977). Accuracy of eye position information for motor control. *Vision Research, 16,* 919-926.

Harris, L. R. (1980). The superior colliculus and movements of the head and eyes in cats. *Journal of Physiology, 300,* 367-392.

Helmholtz, H. von. (1962). *Treatise on physiological optics.* (J. P. C. Southall, Ed. and Trans., Vol. 3) New York: Dover. (Original work published 1867).

Hershberger, W. A. (1976). Afference copy, the closed-loop analogue of von Holst's efference copy. *Cybernetics Forum, 8,* 97-102.

Hershberger, W. (1987). Saccadic eye movements and the perception of visual direction. *Perception & Psychophysics, 41,* 35-44.

Hershberger, W., & Misceo, G. (1983). A conditioned weight illusion: Reafference learning without a correlation store. *Perception & Psychophysics, 33,* 391-398.

Hinde, R. E. (1970). *Animal Behavior* (2nd. ed.). New York: McGraw-Hill.

Honda, H. (1989). Perceptual localization of visual stimuli flashed during saccades. *Perception & Psychophysics, 45,* 162-174.

Howard, I. P. (1982). *Human visual orientation.* New York: John Wiley & Sons.

Jackson, J. H., & Paton, L. (1990). On some abnormalities of ocular movements. *Lancet, 176,* 900-905.

Judge, S. J. (1973). Temporal interaction between human voluntary saccades and rapid phases of optokinetic nystagmus. *Experimental Brain Research, 18,* 114-118.

Keller, E. L., & Robinson, D. A. (1971). Absence of a stretch reflex in extraocular muscles of the monkey. *Journal of Neurophysiology, 34,* 908-919.

Kornmueller, A. E. (1931). Eine experimentelle Anasteise der ausseren Augenmuscheln am Menschen und ihre Auswirkungen. *Journal fur Psychologie und Neurologie, 41,* 354-366.

MacKay, D. M. (1962). Theoretical models of space perception. In C. A. Muses (Ed.), *Aspects of the theory of articificial intelligence.* New York: Plenum.

MacKay, D. M. (1973). Visual stability and voluntary eye movements. In R. Jung (Ed.), *Handbook of sensory physiology* (Vol. 7, pp. 307-332). New York: Springer.

Matin, L. (1972). Eye movements and perceived visual direction. In D. Jameson & L. Hurvich (Eds.), *Handbook of sensory physiology* (Vol. 7, pp. 331-380). Heidelburg: Springer.

Matin, L. (1982). Visual localization and eye movements. In W. A. Wagenaar, A. H. Wertheim, & H. W. Leibowitz (Eds.), *Symposium on the study of motion perception.* New York: Plenum.

Matin, L., Matin, E., & Pearce, D. G. (1969). Visual perception of direction when voluntary saccades occur. I. Relation of visual direction of a fixation target extinguished before a saccade to a flash presented during the saccade. *Perception & Psychophysics, 5,* 65-80.

Matin, L., Matin, E., & Pola, J. (1970). Visual perception of direction when voluntary saccades occur. II. Relation of visual direction of a fixation target extinguished before a saccade to a subsequent test flash presented before the saccade. *Perception & Psychophysics, 8,* 9-14.

Matin, L., & Pearce, D. G. (1965). Visual perception of direction for stimuli flashed during voluntary saccadic eye movements. *Science, 148,* 1485-1488.

Matin, L., Pearce, D. G., Matin, E., & Kibler, G. (1966). Visual perception of direction in the dark: Roles of local sign, eye movements, and ocular proprioception. *Vision Research, 6,* 453-469.

Matin, L., Picoult, E., Stevens, J. K., Edwards, M. W., Young, D., & MacArthur, R. (1982). Oculoparalytic illusion: Visual-field dependent spatial mislocalizations by humans partially paralyzed with curare. *Science, 216,* 198-201.

Matthews, P. B. C. (1972). *Mammalian muscle receptors and their central action.* London: Arnold.

Mays, L. E., & Sparks, D. L. (1980a). Dissociation of visual and saccade-related responses in superior colliculus neurons. *Journal of Neurophysiology, 43,* 207-232.

Ono, H. & Nakamizo, S. (1977). Saccadic eye movements during changes in fixation to stimuli at different distances *Vision Research, 17,* 233-238.

O'Regan, J. K. (1984). Retinal versus extraretinal influences in flash localization during saccadic eye movements in the presence of a visible background. *Perception & Psychophysics, 36,* 1-14.

Park, R. S., & Park, G. E. (1933). The center of ocular rotation in the horizontal plane. *American Journal of Physiology, 104,* 545-552.

Rafal, R., Smith, J., Krantz, J., Cohen, A., & Brennan, C. (1990). Extrageniculate vision in hemianopic humans: Saccade inhibition by signals in the blind field. *Science, 250,* 118-121.

Raphan, T., Cohen, B., & Matsuo, V. (1977). A velocity-storage mechanism is responsible for optokinetic nystagmus (OKN), optokinetic after-nystagmus (OKAN) and vestibular nystagmus. In R. Baker & A. Berthoz (Eds.), *Control of gaze by brain stem neurons* (pp. 37-48). Amsterdam: Elsevier/North-Holland.

Robinson, D. A. (1970). Oculomotor unit behavior in the monkey. *Journal of Neurophysiology, 33,* 393-404.

Robinson, D. A. (1973). Models of the saccadic eye movement control system. Kybernetik, 14, 71-83.

Robinson, D. A. (1975). Oculomotor control signals. In G. Lennerstrand & P. Bach-y-Rita (Eds.). *Basic Mechanisms of ocular motility and their clinical implications* (pp. 337-374). New York: Pergamon Press.

Robinson, D. A. (1981). The use of control systems analysis in the neurophysiology of eye movements. *Annual Review of Neuroscience, 4,* 463-503.

Robinson, D. A. (1986). The systems approach to the oculomotor system. *Vision Research, 26,* 91-99.

Schiller, P. H. (1970). The discharge characteristics of single units in the oculomotor and abducens nuclei of the unanesthetized monkey. *Experimental Brain Research, 10,* 347-362.

Schiller, P. H., True, S. D., & Conway, J. L. (1979). Effects of frontal eye field and superior colliculus ablations on eye movements. *Science, 206,* 590-592.

Schlag, J., & Schlag-Rey, M. (1983). Interface of visual input and oculomotor command for directing the gaze on target. In A. Hein & M. Jeannerod (Eds.) *Spatially oriented behavior* (pp. 87-104). New York: Springer-Verlag.

Schneider, G. E. (1969). Two visual systems. *Science, 163,* 895-902.

Shebilske, W. L. (1977). Visuomotor coordination in visual direction and position constancies. In W. Epstein (Ed.) *Stability and constancy in visual perception.* New York: John Wiley & Sons.

Siebeck, R. (1953). Wahrnehmungsformen bei experiementallen Augenmuskellahmungen. *Deutsche Ophthalmologische Gesellschaft, 58,* 24.

Siebeck, R. (1954). Wahrnehmungsstorung and Storungswahrnehmung bei Augenmuskellahmungen. *von Grafes Archiv fur Ophthalmologie, 155,* 26-34.

Siebeck, R., & Frey, R. (1953). Die Wirkungen muskeleschlaffender Mittel auf die Augenmuskeln. *Anaesthesist, 2,* 139-141.

Skavenski, A. A. (1972). Inflow as a source of extraretinal eye position information. *Vision Research, 12,* 221-229.

Skavenski, A. A., Haddad, G., & Steinman, R. M. (1972). The extraretinal signal for the visual perception of direction. *Perception & Psychophysics, 11,* 287-290.

Sparks, D. L. (1986). Translation of sensory signals into commands for control of saccadic eye movements: Role of primate superior colliculus. *Psychological Reviews, 66,* 118-171.

Sparks, D. L., & Mays, L. E. (1983a). Role of the monkey superior colliculus in the spatial localization of saccadic targets. In A. Hein & M. Jeannerod (Eds.) *Spatially oriented behavior* (pp. 63-86). New York: Springer-Verlag.

Sparks, D. L., & Mays, L. E. (1983b). Spatial localization of saccadic targets. I. Compensations in eye position. *Journal of Neurophysiology, 49,* 45-63.

Sparks, D. L. Mays, L. E., & Pollack, J. C. (1977). Saccade-related unit activity in the monkey superior colliculus. In R. Baker & A. Berthoz (Eds.), *Control of gaze by brain stem neurons* (pp. 437-443). Amsterdam: Elsevier/North-Holland.

Sparks, D. L., & Porter, J. D. (1983). Spatial localization of saccade targets. II. Activity of superior colliculus neurons preceding compensatory saccades. *Journal of Neurophysiology, 49,* 64-74.

Sperry, R. W. (1950). Neural basis of the spontaneous optokinetic response produced by visual neural inversion. *Journal of Comparative & Physiological Psychology, 43*, 482-489.

Stark, L., & Bridgeman, B. (1983). Role of corollary discharge in space constancy. *Perception & Psychophysics, 34,* 371-380.

Steinbach, M. J. (1986). Inflow as a long-term calibration of eye position in humans. *Acta Psychologica, 63*, 297-306.

Stevens, J. K., Emerson, R. C., Gerstein, G. L., Kallos, T., Neufeld, G. R., Nichols, C. W., & Rosenquist, A. C. (1976). Paralysis of the awake human: Visual perceptions. *Vision Research, 16*, 93-98.

Trevarthen, C. B. (1968). Two mechanisms of vision in primates. *Psychologische Forschung, 31*, 299-337.

Turvey, M. T., & Carello, C. (1986). The ecological approach to perceiving-acting: A pictorial essay. *Acta Psychologica, 63*, 133-155.

Ullman, S. (1980). Against direct perception. *Behavioral and Brain Sciences, 3*, 373-415.

von Graefe, A. (1854). Beitrage zur Physiologie und Pathologie der schiefen Augenmuskeln. *Graefes Archiv fur Ophthalmologie, 1*, 1-18.

von Holst, E., & Mittelstaedt, H. (1950). Das Reafferenzprinzip. *Naturwissenshaften, 37*, 464-476.

Wallach, H. (1987). Perceiving a stable environment when one moves. *Annual Review of Psychology, 38*, 1-27.

West, R. (1932). Curare in man. *Royal Society of Medicine, Proceedings, 25*, 1107-1116.

Woodworth, R. S., & Schlosberg, H. (1954). *Experimental psychology*. New York: Holt, Rinehart, & Winston.

Wurtz, R. H., Goldberg, M. E. (1971). Superior colliculus cell responses related to eye movements in awake monkeys. *Science, 171*, 82-84.

The Role of Eye Movements in Perceptual Processes
E. Chekaluk and K.R. Llewellyn (Editors)

CONTEXTUAL FACTORS IN SCENE PERCEPTION

Helene Intraub

1. Introduction

A picture depicts only a part of a real-world scene. The artist determines how much of the scene will be shown within the picture's boundaries. Comprehension of the picture, however, cannot be limited to the depicted information alone, but must take into account the context that "exists" just outside the picture's boundaries. For example, when we study the portrait of a friend, we see only a head and neck, yet we don't gasp at the apparent decapitation of a comrade. In perceiving the portrait, we perceive that the rest of the friend and the rest of the scene continue beyond the edges of the picture. It is as if we were viewing the person through a window. In a sense, the structure of the world just outside the boundaries of a picture may be as tangible to the perceiver as information that exists just outside a window frame.

A bounded picture in many ways may be considered analogous to the "visual picture" captured by a single eye fixation during visual scanning. An eye fixation provides the viewer with a detailed glimpse of only a part of the available scene. Visual acuity drops off rapidly for information falling outside the fovea on each fixation. Although, unlike the case of the picture, the information in the periphery is not excluded (or occluded as in the case of the window), it is degraded. This means that only a part of the visual field can be perceived clearly at any given time. This characteristic of visual processing is the basis of a classic question in the field of perception. How can successive, spatially constrained, discrete eye fixations yield perception of a stable, continuous, visual world. The level at which these successive fixations are integrated with one another, thus yielding the perception of a continuous visual world, remains in question.

It has been acknowledged that retinotopic visual persistence (a retinotopic icon) is not a likely medium for the integration of eye fixations (e.g., Haber, 1983; Neisser, 1967). Because the viewer makes numerous head and eye movements during visual processing, information from each successive fixation would be likely to overlap the same retinal area, resulting in more interference (through masking) than integration across saccades. Spatiotopic visual persistence, however, would eliminate this problem. It has been suggested that integration of successive iconic representations based upon spatial rather then retinal coordinates might provide the viewer with the perception of a stable, continuous visual world.

The classic research associated with this hypothesis was conducted by Davidson, Fox, and Dick (1973). They presented subjects with a display of letters during one fixation and then had them shift fixation slightly to a new location. During this new fixation a visual mask was presented at one of the letter positions. They reported that the letter that had shared *retinal* coordinates with the mask, was masked but the letter that had shared its *spatial* coordinates was perceptually integrated with the mask. This intriguing observation suggested that the stable visual world we perceive might be the result of spatial persistence and integration in a short-term visual memory. However, there have been some criticisms of this research on methodological grounds coupled with a failure to demonstrate spatiotopic persistence and integration when methodological changes were introduced (e.g., Irwin, Yantis, & Jonides 1983; Irwin, Brown, & Sun, 1988).

The possibility that successively glimpsed visual areas persist and are pieced together in an integrative memory is still open. However, another possibility is that visual persistence is not the primary medium that underlies visual comprehension of successive fixations. It is possible that each fixation is perceived in relation to an abstract mental representation of the environment that serves as a mental map (e.g., Hochberg, 1978, 1986; Irwin, Brown, & Sun, 1988). It is this mental map that provides the visual context within which each successively viewed area is "placed" and understood.

Bearing this in mind, consider once again the analogy between a picture and a single eye fixation on a scene. If a mental schema that represents the area outside the boundaries of a picture plays an integral role in picture perception, this same representation may play a role in the perception of successive eye fixations during visual scanning. The schema would provide a storage system for the incorporation of currently fixated information with previously fixated information, and would provide visual/spatial expectations about the information that is likely to be extracted in subsequent fixations. The same mental map that allows us to understand a close-up of a friend's face may underlie our ability to rapidly understand successive views.

How to study the nature and, in fact, the existence of such a schema is a difficult question. In this chapter I will provide a description of a picture memory phenomenon called "boundary extension" (Intraub and Richardson, 1989), that may provide a means for studying the perceptual schema used in picture perception. The basic observation is that when recollecting a picture, observers tend to remember having seen information that was not present in the picture itself, but that is likely to have existed just outside the picture's boundaries. Using recall and recognition tests, my students and I have been attempting to determine if the "additional" information is derived from the activation of a mental schema during picture perception. Recent research designed to test this hypothesis will be reported. Finally, the implications that this research and other picture processing research have for our understanding of the integration of successive fixations during visual scanning will be addressed.

I will begin with a review of Hochberg's (1978, 1986) analysis of the role of mental schemata or maps in perception, as this will provide the basis for one of the major hypotheses tested in the boundary extension experiments.

2. The Perceptual Schema

Hochberg (1978, 1986) has been a major proponent of the view that an abstract, schematic, mental structure plays a fundamental role in perception. Hochberg has argued convincingly that such a structure must underlie the comprehension of successive glimpses, basing his case upon an analysis of the viewer's ability to understand motion picture and video displays. Such displays include rapid shifts of the camera's view - shifts that the viewer could never make via his or her own locomotion. Hochberg argues that the ease with which an observer comprehends these shifts in viewpoint raises serious problems for the theory of direct perception (e.g., Gibson, 1950). Furthermore, he proposes that the capabilities suggest the use of mental schemata of space and events within which successive views are analyzed and understood. He demonstrates the likelihood of such a mental structure through an analysis of film editing techniques and descriptions of laboratory experiments. One such experiment is particularly relevant to the discussion here.

Hochberg (1978, 1986) describes research in which a movie was made which showed a circular aperture through which successive nonoverlapping views of an outline cross could be seen. Each view was a close-up of a part of the cross, so that only two perpendicular lines were visible at a time. The views were changed at a rate of 2-3 per second. When viewers were unaware that the display was intended to depict an outline cross moving behind an aperture, they often perceived it as hands moving around the circular face of a clock. However, when the viewer was told about the cross behind the aperture, or was provided with a long shot of the cross followed by a medium shot and close-up, thus establishing that the cross was behind the aperture, the perception changed. Although the aperture never allowed the whole outline cross to be seen at one time, the viewers "perceived" its existence outside the boundaries of the circle. Their perception of a cross moving behind a circular opening was accurate enough to allow them to recognize when one of the arms of the cross had been skipped.

The point of this demonstration is that the viewer's visual perception depended not only on the stimulus input, but upon a mental structure he or she brought to the event. However, the viewer's awareness of the "aperture" interpretation of the display alone, was not in itself responsible for the effect. Given the same description of the display, perception of the moving cross did not occur when the sequence of successive views was presented more rapidly (e.g., 10 views/second). In this case visual integration occurred, causing the successive views to summate. The demonstration was successful only when presentation rates of at least 2-3 views/second were used (Hochberg, 1986). The slower rates are more consistent with the rate at which viewers shift their fixation during visual scanning (e.g., Yarbus, 1967). At these rates, sensory/perceptual integration was not evident, and viewers apparently were able to make use of the proposed mental schema.

This demonstration illustrates the primary difference between the mental schema proposed by Hochberg (1978, 1986), and the type of visual representation proposed by Davidson, Fox, and Dick (1973). The schema is not considered to be sensory or photographic in nature. Instead, successive views are thought to be stored in a more abstract format. Hochberg (1986) points out that during visual scanning, many visual details may simply go unnoticed and may be unrepresented in the schema. He illustrates this point with a scene from the movie, "Nights of Cabiria". In a two-shot series of close-ups, a truck which was visible over the shoulder of an actor disappeared from one cut to the next -- a "disappearing act" that viewers of the film simply do not notice. The mental schema maintains important spatial and form-related information, without being a sensory representation.

To summarize, the mental schema is considered to be a part of the perceptual process. This schema provides the viewer with the means for comprehending successive views, and allows the viewer to grasp the structure of a visual world that is never visually present all at once, but "exists" beyond the boundaries of each successive view. In visual scanning, the schema would provide expectations about the probable layout of the next eye fixation, as well as providing a spatially organized storage system for incorporating information gleaned from successive glimpses of the world. This structure is so important to the final perception, that Hochberg (1986) argues that "... the schematic events that the viewer has in mind and can bring to the moving picture are, in normal usage, at least as important as the stimulus information, regardless of how sophisticated the analysis of the latter" (p 48).

The perceptual schema, however, is difficult to study. Hochberg's analysis of film and video editing provides one means for inferring its nature. Another possible means of studying the perceptual schema may come from a recently reported picture memory phenomenon called, "boundary extension".

3. Boundary Extension

What is the mental representation of a picture like? How well do people actually recollect the details and the layout of a picture? The possibility that drawings might provide insight into pictorial representation led me to study students' drawings of previously viewed photographs. Although their artistic competence varied widely, I began to notice that their drawings all seemed to contain the same error. In photographic terms, their drawings depicted a more wide-angle view of the scene than had been captured in the photograph. Put another way, nearly all the drawings contained visual information that is likely to have existed just outside the camera's field of view, but that had not been present in the photograph they had studied. I refer to this error as "boundary extension", because the subjects had apparently extended the boundaries of the picture. The question of why this error was the rule rather than the exception was intriguing, but the most interesting aspect of the phenomenon for me was the possibility that memory of a picture might include information drawn from a perceptual schema that was active during picture perception.

Intraub & Richardson (1989) recently conducted a formal test of this observation. First, we replicated it using drawing tasks, then we demonstrated that the same memory error was evident in recognition tests as well. In these experiments, we tested memory for close-up views of natural scenes following relatively long retention intervals (35 minutes or 2 days). The results clearly showed a unidirectional distortion of picture boundaries in recall and recognition of previously viewed photographs.

In one experiment, 37 subjects were presented with close-up photographs (35 mm slides) of 20 common scenes for 15 seconds each. In all the scenes the main objects had been cropped by at least one of the picture's edges (for an example, see Figure 1, panel A). They were instructed to remember each picture in as much detail as possible, and to consider the background to be as important as the foreground. After a 35 minute retention interval, subjects were provided with response sheets containing rectangles bearing the same aspect ratio as a 35 mm slide (1:1.5). They were instructed to draw four of the pictures. Each picture was to be drawn within one of the 4" x 6" rectangles. They were told to consider the edges of the rectangle to be the edges of the photograph and to draw the pictures accordingly.

Although all of the photographs contained main objects that were cropped, subjects did not depict them in this way in their drawings. Instead, they completed the main object(s) and added more background information between them and the picture's edges than had actually been the case. Using conservative criteria (see Intraub and Richardson, 1989), judges rated each drawing. Out of 133 drawings, 95% clearly showed boundary extension. The remaining 5% were either rated as accurate or as showing boundary restriction.

A second experiment was carried out to determine if completion of the cropped main objects was causing the distortion. Two versions of each scene were prepared: One in which the main objects were slightly cropped, and one in which they were not (see the photographs in Figure 1). Forty-one subjects were presented with the same 20 scenes for 15 seconds each. Half were shown in their cropped version and half were shown in their slightly wider-angle uncropped version. Across subjects, scenes were presented in each version equally often. After 48 hours, subjects were asked to draw up to 6 of the pictures. Again, they were instructed to lay out the picture in the rectangle so that the edges were the same as in the stimulus. The results showed that regardless of whether or not the main objects were cropped, subjects tended to extend the boundaries of the stimuli. Boundary extension was evident in 96% of the drawings associated with the cropped stimulus versions, and in 87% of the drawings associated with the wider-angle stimulus versions.

Two representative drawings are shown in Figure 1. To evaluate the drawings, it is important to look at the pictures' edges -- first at the photograph and then at the drawing. It is interesting to note that the typical drawing of the cropped close-up (panel C) looks like the wider-angle photograph (panel B) that the subject did not see. The typical drawing of the wider-angle photograph (panel D), is an even more wide-angle depiction. Apparently, subjects were not simply drawing an uncropped view of the main objects. Whatever boundaries the edges of the photograph had imposed, subjects depicted them as having revealed more of the scene than had actually been the case.

Intraub and Richardson (1989) demonstrated that this unidirectional distortion was not an artifact of the drawing task. The same distortion was evident in a recognition test in which subjects rated test pictures (targets and distractors) in terms of how similar their boundaries were to the boundaries of the presentation pictures. In this experiment, pairs of scenes such as the example in Figure 1 were used in the following way.

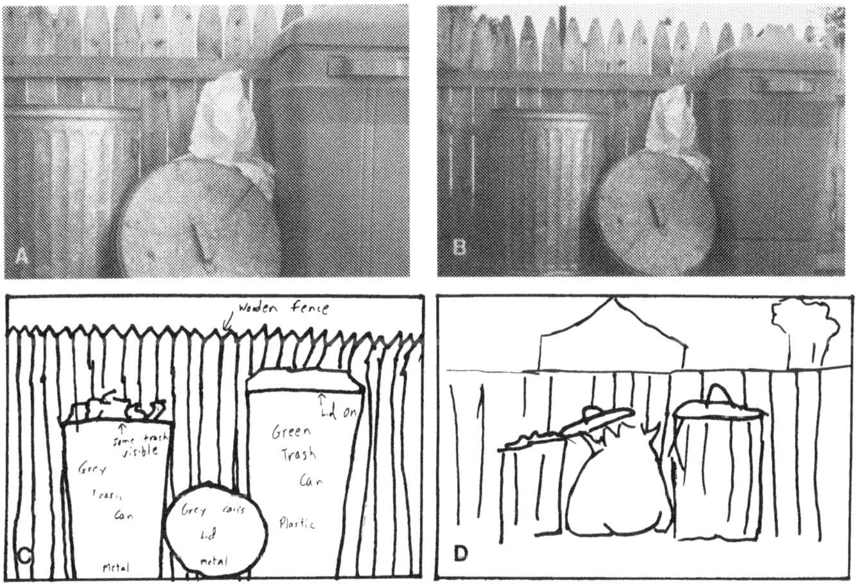

Fig. 1. Scene pair and representative drawings showing boundary extension. Panel A shows a close-up version of the scene with main objects cropped. Panel B shows a slightly more wide-angle version of the same scene with main objects uncropped. The drawings in Panels C and D are representative drawings of the photographs in Panels A and B, respectively. The actual photographs were in color. The subjects' original drawings were in pencil; these were photocopied and all lines darkened for this figure. (From Intraub & Richardson (1989), copyright 1989 by the American Psychological Association. Adapted by permission.)

Eighty-five subjects viewed 20 photographs of scenes for 15 seconds each: 10 in their relatively close version, and 10 in their slightly wider-angle version. Two days later, subjects were presented with the same 20 scenes in the same order of presentation. Half the scenes were presented in the original version and half were presented in the other version. They were asked to rate each test picture on a 5-point scale as "a lot closer-up" (-2), "a little closer up" (-1), "the same" (0), "a little further away (+1)", or "a lot further away" (+2).

Subjects' responses showed that their pictorial representations of the pictures contained extended boundaries. They tended to rate the same pictures as being "closer-up" than before. This response to seeing the *same* picture again, indicated that the subject's representation of that picture had extended boundaries. Subjects made significantly more boundary extension responses than boundary restriction responses. This was particularly pronounced for the closer views which were rated as closer than before 65% of the time, and wider than before only 3% of the time. The average rating on the 5-point scale was -.68 for the closer pictures, and -.12 for the wider-angle pictures. The pattern of responses elicited by the distractors also indicated that subjects were remembering the pictures with extended boundaries.

Given two versions of the same scene, an asymmetry was observed depending on whether the closer version was the stimulus and the slightly wider version was the distractor, or vice versa. Once again, refer to the stimulus pair in Figure 1 (panels A and B). When the wider-angle version was the distractor, it was rated as closer to "same" (.24) than when the closer version was the distractor (-1.30). This asymmetry is the logical result of a unidirectional distortion. If the presentation picture is remembered with extended boundaries, a wider-angle distractor would be expected to more closely match the subject's recollection than a distractor showing a closer view. The same pair of pictures were responded to differently depending on which was remembered and which served as the test item.

The key observations in these experiments are: 1) Subjects did not simply exhibit poor memory for picture boundaries -- sometimes calling a picture a little closer up and sometimes calling it a little further away. Their errors tended to be unidirectional, yielding boundary extension. 2) Both versions of the scenes used in these experiments can be described as close-ups and both versions yielded boundary extension. The closer versions, however, yielded a greater degree of extension than the slightly wider versions. This was the case in the recognition test and was also reflected in the recall (drawing) task.

The nature of the distortion is intriguing in the light of the discussion of the perceptual schema. One hypothesis is that when perceiving the picture, a schema, extending beyond the picture's boundaries, is activated during comprehension of the close-up view of a natural scene. This schema is so important to the comprehension of the picture, that it becomes incorporated into the subject's recollection of the picture. Highly probable parts of the schema are remembered as having been physically present. There are however, two other strong alternate explanations, one having to do with normalization processes in memory (memory schema), and one having to do with the completion of background objects.

4. Boundary Extension and the Perceptual Schema

Three alternative explanations of boundary extension will be discussed in this section: the perceptual schema hypothesis, the memory schema hypothesis, and object completion.

Perceptual Schema Hypothesis. The perceptual schema hypothesis is predicated on two assumptions. The first is that picture perception involves the activation of a perceptual schema that extends beyond the picture's boundaries. The second is that perceptual expectancies contained in that schema may become incorporated into the subject's recollection of the picture. Consistent with the view that the mental schema is as important to perception as the actual stimulus input, in memory the observer cannot distinguish between the two. In addition to the depicted information, the subject recollects having seen the highly probable information that he or she had previously "perceived" as existing just outside the picture's boundaries.

Intraub and Richardson's (1989) observation that slightly wider-angle pictures yielded less extension than the closer views can be explained by this hypothesis. Consider several views of a centrally located object, ranging from a close-up view to a wide-angle view. In the case of a tight close-up, highly probable surrounding information will not be captured in the photograph. It will, however, be represented in the subject's mental schema during perception of the picture and in fact is critical to the observer's comprehension. As more wide-angle views of the same object are presented, more of the probable surrounding area will be contained within the picture itself.

As a result, boundary extension would be greatest for close-ups and would decrease for increasingly wide-angle views. At some point, a view might become wide enough for the amount of extension to asymptote, and in fact there may be a point at which no *directional* error would be obtained. Subjects would sometimes consider the picture to be "just a little closer up than before" and sometimes, "just a little further away." This model would predict overall boundary *restriction* for none of the picture types. Put another way, if a picture can be thought of as analogous to the information in a single eye fixation, then comprehension of the picture may include the expectation of what the next eye fixation would be likely to bring into view. This means that whenever there is a directional distortion it should move the boundaries outward -- not inward to yield restriction.

Memory Schema Hypothesis. According to the memory schema hypothesis, boundary extension reflects a process of normalization in memory that can be thought of as "regression to the prototype" (cf. Bartlett, 1932; E. Gibson, 1969). Subjects may have an expectation about a standard view of an object. The close-ups used in Intraub and Richardson's (1989) experiments, are by definition closer than "the standard view". Over time, the subject's representation becomes biased toward the prototypic viewing distance. As a result, a close-up is remembered as having been less of a close-up.

This hypothesis provides a different explanation of why the wider versions yielded less extension than the closer versions in Intraub and Richardson's (1989) experiments. The wider versions were closer to being prototypic to begin with, and therefore underwent a less dramatic transformation as they normalized. Following this logic, the memory schema hypothesis leads to a different prediction than the perceptual schema hypothesis, regarding the presentation of increasingly wide-angle views.

If we begin with close-ups, then, according to this hypothesis, as increasingly wide-angle views are presented subjects should show decreasing boundary extension, reaching no directional distortion for prototypic pictures, and then reversing toward increasing degrees of boundary *restriction* as more wide-angle views are presented.

To summarize, the perceptual schema hypothesis attributes boundary extension to the initial comprehension of a picture which includes schematic expectations about information just outside the picture's boundaries. Alternatively, the memory schema hypothesis attributes the phenomenon to normalization of the representation toward a prototypic viewing distance.

Each can account for the results of Intraub and Richardson (1989), but each predicts a very different pattern of results regarding memory for prototypic and wide-angle pictures.

Object Completion. In the wider-angle versions of the pictures used by Intraub and Richardson (1989), although the main objects were not cropped by the picture's edges, the edges frequently cropped background objects, such as the door to a room or a window in the background, etc. For this reason, the Gestalt principle of object completion (cf. Ellis, 1955) was considered as a possible explanation of boundary extension. Intraub and Richardson (1989) tentatively argued against object completion as the cause of boundary extension based upon some characteristics of their subjects' drawings (e.g., subjects frequently extended boundaries without completing the cropped objects -- they simply drew more of the object than had appeared in the original picture). To provide a direct test of the hypothesis, however, it would be important to determine if boundary extension would occur if there were no cropped background objects. All of the following experiments made use of new stimuli that met this requirement. If boundary extension is caused by a tendency to complete background objects, then no boundary extension should be obtained with these new pictures.

5. New Boundary Extension Research

Intraub, Bender, and Mangels (1992) conducted a series of experiments to determine whether boundary extension reflects object completion, normalization to an expected viewing distance, or the activity of a perceptual schema during picture perception. To test these three hypotheses, a new stimulus set was created that will be described shortly. In addressing the two schema hypotheses, Intraub, et. al. (1992) considered the possibility that both the perceptual schema and the memory schema might affect pictorial representation, but that these effects might follow different time courses. Specifically, the effects of a perceptual schema might be more readily apparent immediately following presentation, whereas the effects of a memory schema (normalization toward a prototypic viewing distance) might take place over a longer period of time. To test this possibility, boundary memory was tested immediately as well as after a 2-day delay.

To create a stimulus set that would allow a test of the three hypotheses, 22 scenes were photographed in three different versions: close-up, medium, and wide-angle. The pictures always contained a main object (or objects) against a natural, textured background that contained no incomplete objects (e.g., asphalt, brick, grass, carpeting, tile, etc.). Fifty-eight subjects rated the pictures on a 5-point scale that ranged from -2 to +2, to indicate if a picture depicted a standard view of the object (0), was too close-up to be standard (-1 or -2, depending on degree), or was too far away to be standard (+1 or +2). Each subject was presented with one version of each of the 22 scenes. Sixteen of the scenes were selected for the new stimulus set based on the group ratings. The mean ratings for close-ups, prototypes, and wide-angle pictures in this set were -.42, -.02, and +1.53, respectively. Although the photographs in Figure 2 are from a more recent picture set (to be described later), they are similar to the scenes used by Intraub et al. (1992), and will give a good indication of what the stimuli looked like.

Intraub et al. (1992) replicated the boundary extension effect with close-ups from the new stimulus set using the same procedure as in the previous research. Although the new scenes were more simple than those presented by Intraub and Richardson (1989), the new close-ups yielded a comparable degree of boundary extension. Across experiments, the new stimulus set yielded no support for the object completion hypothesis. Boundary extension was readily apparent in memory for pictures that had contained no cropped background objects. Subjects continued to rate the same view as being "closer-up" than before, even though the backgrounds were homogeneous. They simply recollected having seen more of the background.

A test of the predictions of the two schema hypotheses was conducted in two experiments, each including both an immediate and 2-day delay condition. In the first of the immediate/delay experiments, we presented subjects with prototypic and wide-angle pictures, using the same basic design as Intraub and Richardson (1989). Subjects in this experiment viewed 16 scenes (half in their prototypic version and half in their wide-angle version). Half the subjects were tested immediately after presentation, and half were tested after a 2-day delay. In the recognition test half of the pictures were presented in the original version (targets) and half in the other version (distractors). Subjects rated each picture on the same 5-point scale described previously.

Boundary extension was evident within minutes of presentation. When memory was tested immediately, the pattern of results clearly supported the perceptual schema hypothesis; prototypes were remembered with extended boundaries (mean rating = -.13), and the wide-angle pictures yielded no directional distortion (mean rating = -.03). It is important to note that the wide-angle pictures were *not* remembered more accurately than the prototypes - the hit rate (correctly rating the pictures as "same") for the two versions did not differ. What differed was that when subjects saw the prototypes again, they made significantly more extension errors than restriction errors, whereas when they saw the same wide-angle picture again, their errors were equally distributed, yielding no directional bias.

When memory was tested after a two day delay, a different pattern of results was obtained. Consistent with the memory schema hypothesis, wide-angle pictures were remembered with *restricted* boundaries - yielding a mean rating of +.32. Upon viewing the same wide-angle pictures in the test, subjects made significantly more restriction responses than extension responses. They tended to rate the wide-angle pictures as being "further away" than before, thus indicating that they remembered those pictures as having had more restricted boundaries. However, the results only partially supported the memory schema hypothesis. Contrary to one of the predictions, the prototypic pictures yielded boundary extension. In fact, boundary extension for the prototypes increased over time, yielding a mean score of -.25. One possibility was that in spite of what was indicated by the normative ratings, the prototypes might not actually have been prototypic. Another possibility, however, was that the memory schema hypothesis required modification.

A second experiment was conducted in order to replicate and extend the results of the immediate condition, as well as to explore an alternative memory schema hypothesis. The alternate hypothesis was suggested by the symmetry of the delay results. Perhaps, over time, the pictorial representations normalize toward the average view depicted in the stimulus set, rather than toward a prototypic viewing distance. That is, the pictorial representations may have normalized toward an average of the prototypic and wide-angle views presented. In this experiment, therefore to avoid the averaging of extremely different views, subjects viewed only one picture type: close-up, prototypic, or wide-angle. The question was whether boundary extension would be observed for relatively homogeneous picture sets. Of particular interest, given the memory schema hypothesis, was whether it would occur for prototypes under these conditions.

One hundred and thirty subjects were divided into three groups. Each group was presented with either the close-up, prototypic, or wide angle version of all 16 scenes for 15 seconds each. Half of the subjects were tested immediately, and half were tested after a two-day delay. To avoid any contamination of memory through the introduction of other versions at any point in the experiment, the recognition test sequence contained the same 16 pictures as the presentation sequence with no distractor items. Subjects were told that the same scenes would be presented but that sometimes they might be slightly more wide-angle or slighter more close-up versions. They were instructed to rate each picture on the same 5-point scale used in the prior research to indicate if the pictures were the same as before (0), closer up than before or more wide-angle than before. If the prototypes in the previous experiment had yielded boundary extension because they were mixed with wide-angle pictures, then they should show no boundary extension in an experiment in which they are presented alone.

The immediate condition provided clear support for the perceptual schema hypothesis: when different picture types were not mixed together, boundary extension was obtained for close-ups, prototypes, and wide-angle views. The results were all the more striking because the subjects were viewing the *same* 16 pictures within minutes of their initial presentation. In all three conditions, subjects made significantly more boundary extension errors than restriction errors. Furthermore, the degree of extension decreased significantly as increasingly wide-angle views were presented. The mean ratings obtained for each picture type at both retention intervals are presented in Table 1.

The decrease in boundary extension with increasingly wide-angle views could not be attributed to better memory for the wider-angle versions, because as in the previous experiment, the hit rate (number of pictures correctly identified as "same") did not differ among the three conditions. When tested immediately, subjects' recollections of pictures tended to have extended boundaries and did not seem to be influenced by normalization to an expected view. Contrary to the memory schema hypothesis, none of the picture types yielded boundary restriction, and pictures in the prototypic range yielded boundary extension.

As in the previous experiment, a 2-day delay resulted in a different pattern of errors. As shown in Table 1, following a delay, wide-angle pictures now yielded boundary restriction. The directional distortion was small but significant. Unlike the previous experiment (in which picture versions were mixed), all three picture types showed a decrease in extension over time. This was not due to increased accuracy after a delay, but to an increase in the number of restriction responses. This "restrictive shift" occurred for all three stimulus types and led to a small but significant bias toward restriction for the wide-angle pictures.

Table. 1. Mean Boundary Ratings for Close, Prototypic, and Wide-Angle Pictures as a Function of Retention Interval when Picture Versions were Unmixed

	Picture Version		
Retention Interval	Close	Prototype	Wide
Immediate	-.45	-.34	-.17
Delay	-.28	-.20	.07

The pattern of results in these experiments suggests a two-component model of the representation of pictorial views. Shortly after studying pictures, observers remember them with extended boundaries. According to the model, this unidirectional distortion of the boundaries reflects the activation of a perceptual schema during picture comprehension. Highly probable scene information that was not visually present is included in the subject's recollection of the picture. Over time, however, the representation undergoes a transformation in memory. Contrary to the memory schema hypothesis, this transformation does not appear to include normalization to an expected viewing distance for the objects. What it may include is an interaction of the early effects of the perceptual schema (which tends to move the boundaries out) with a tendency for the pictures to normalize toward the average of the set (even within a very similar set, such as that used in the present experiment).

This latter view was supported by the results of a condition in which the same pictures were presented with slightly more wide-angle versions of other pictures, in a set that on average was rated as prototypic, or were presented with slightly more close-up versions of other pictures, in a set that on average was rated as close-up. The results showed that although the pictures all yielded extension, they yielded a greater degree of extension when presented with slightly wider pictures than with slightly closer pictures. Future experiments will analyze the changes in boundary distortion over a wider range of retention intervals to provide a better understanding of the nature of the transformation in memory. The most interesting outcome of the research for the present discussion, however, is the support for the perceptual schema hypothesis in the immediate conditions. Within minutes of viewing photographs subjects remembered them with extended boundaries and the size of the boundary effect was associated with the picture view.

In ongoing research with Deborah Berkowits, we have continued to explore immediate memory for picture boundaries. We have just completed an experiment designed to replicate the immediate results of Intraub et al. (1992) in recall using a drawing task. Two questions of primary interest were whether observers would draw extended boundaries for close-ups, prototypes and wide-angle pictures within minutes of viewing them and whether this would happen if a small stimulus set of only 7 items was presented. Of secondary interest was the introduction of a new stimulus set in which a more conservative estimate of prototypic pictures was used, as a final replication of the observation that prototypic pictures do indeed yield extended boundaries.

Out of a new set of 42 scenes (with 5 versions each), seven scenes were selected that exhibited high subject agreement on which version was prototypic, too close-up to be prototypic and too wide-angle to be prototypic. Once again, no cropped background objects were included in the pictures. The mean boundary ratings for the close-up and wide-angle versions were -1.10 and +1.10, respectively, and the mean rating for the prototypes was -.08. These "prototypes" had yielded higher subject agreement than in the previous experiment (mean agreement = 80%). The photographs in Figure 2 are from this stimulus set, and show the three versions of the "basketball scene".

Subjects were presented with either 7 close-ups, 7 prototypes, or 7 wide-angle pictures for 15 seconds each. As in the earlier experiments, they were told to make an exact copy of the picture in memory and to remember the foreground and the background in as much detail as possible. Immediately following presentation, they were asked to draw each of the seven stimuli. A booklet containing 4" x 6" rectangles was provided, along with the name of each scene in the order of presentation viewed by that subject. After drawing the pictures, the subjects were presented with the same stimuli again in a recognition test and rated each of the 7 pictures on the same 5-point scale used previously. There were 49-50 subjects in each of the 3 conditions (close-up, prototype, wide-angle) who participated in groups of up to 6 at a time.

The drawings provided dramatic support for the perceptual schema hypothesis. Immediately after viewing 7 slides for 15 seconds each, subjects' drawings revealed pronounced boundary extension for the close-ups *and* the prototypes, and revealed no directional distortion for the wide-angle pictures. Sample drawings are presented in Figure 2.

Instead of judges evaluating each drawing with respect to a set of criteria, a quantitative measure of the distortion was evaluated. The area of each main object drawn, was estimated by tracing the object onto graph paper (10 boxes to the inch) and counting the boxes and partial boxes on the grid. The area of the stimulus object was determined by projecting the stimulus slide onto the same grid with the same dimensions as the drawings (4" x 6"). The mean proportion of the stimulus object's area drawn by the subjects for each picture, in each of the three conditions, is presented in Table 2. On average, subjects extended the boundaries of the close-up views such that the main object took up only about 1/3 of the area taken by the stimulus object in the photograph. For prototypes, the drawn object took up only about 4/10 of the stimulus object. In the case of wide angles, no directional distortion was

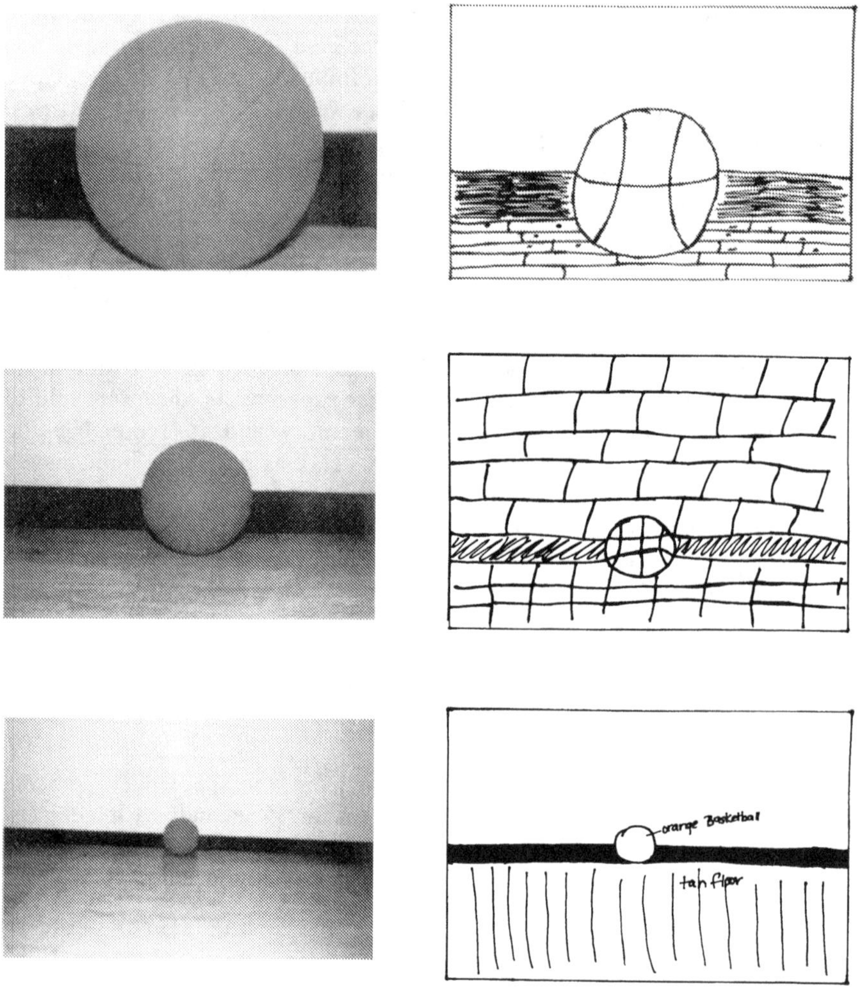

Fig. 2. Three versions of the "basketball scene" with a representative drawing for each. From top to bottom, the figure shows the close-up and prototypic versions (with representative drawings showing boundary extension), and the wide-angle version (with a representative drawing showing no directional distortion.) The actual photographs were in color. Subjects original drawings were in pencil; these were photocopied and all lines darkened for this figure.

obtained for the set. On average, the drawn object and the stimulus object covered the same amount of area in the picture space - the relative proportion was exactly 1.00.

Table 2. The Average Proportion of the Stimulus Object Depicted in the Drawing for each Version (Close-Up, Prototype, and Wide-Angle) of each of the Seven Scenes

	Scene						
Version	Basketball	Bear	Bananas	Tire	Sneakers	Pail	Crayons
Close-up	.43	.34	.26	.44	.36	.41	.22
Prototype	.66	.29	.30	.44	.39	.37	.42
Wide-angle	1.98	1.24	.78	1.20	.81	.68	.53

In the close-up condition, as may be seen in Table 2, each of the seven pictures showed considerable boundary extension. Out of 343 drawings, made by 49 subjects, only 4 drawings depicted an object that matched or exceeded the size of the original. In all the close-up photographs, the main objects had either touched or been slightly cropped by at least one of the picture's edges, yet subjects drew the objects as uncropped, and depicted a considerably more wide-angle view of each scene. Furthermore, subjects obviously had paid attention to the background, because they correctly drew the background that had immediatly surrounded the object -- their error was in drawing a greater expanse of background than had actually been the case.

In accordance with the perceptual schema hypothesis, the prototypic pictures were also remembered with extended boundaries, and the effect was slightly less pronounced than for the close-ups. Out of 350 drawings made by 50 subjects, all but 13 had extended boundaries. The drawings contradicted the memory schema hypothesis' prediction that pictures that already represent the prototypic view should yield no directional distortion; recall that on average the boundaries were extended enough to shrink the main object to 40% of its actual size in the picture space. Comparison of the drawings in the close-up and prototype conditions shows that subjects were not simply drawing an idealized version of the scene. If this had been the case, one would expect to see similar drawings regardless of whether the subject had seen the close-up version or the prototypic version. Instead, whatever boundaries the stimulus picture had placed on the scene, the subjects seemed to extend those boundaries further.

Subjects' drawings of wide-angle pictures yielded no directional distortion. This does not mean that subjects were accurate on each picture. Unlike the other two picture versions, there were many instances of restriction as well as extension for all the pictures. If anything there was a slight bias toward extension, in that four of the pictures tended to be drawn with extended boundaries and three with restricted boundaries (see Table 2). An interesting observation was that the stimuli containing main objects that covered the smallest amount of picture space tended to be the ones yielding restriction, whereas the stimuli with main objects that covered a larger area of the picture space tended to yield extension. (This may be seen in Table 2 in which the pictures are listed in descending size order of the main objects in the wide-angle condition). This observation supports Intraub et al.'s (1992) suggestion that when the effect of the perceptual schema is not strong, the pictures will tend to normalize toward the average of the picture set.

The drawing task took subjects approximately 20 minutes to complete. Following this they were given the same recognition test as in the unmixed condition of the Intraub et. al. (1992) experiment. All seven pictures were presented again and subjects rated each one on the 5-point scale. The results replicated those of the earlier experiment. Subjects made significantly more extension than restriction responses for all three stimulus types (Wilcoxon tests, $p < .001$, all cases) and the mean extension rating decreased significantly as increasingly wide-angle views were presented, $F(2,151) = 14.35$, $MSE = .12$, $p < .001$. Mean boundary ratings were -.66, -.40, and -.30, for the close-up, prototypic, and wide-angle pictures, respectively.

These results and the results of the immediate tests conducted by Intraub et al. (1992), all support the perceptual schema hypothesis. They support the view that boundary extension is the result of the activation of a scene schema during picture perception -- a schema that represents the likely structure of the scene that "exists" just beyond the picture's boundaries. It is this schema, in conjunction with visual analysis of the stimulus, that allows us to perceive a picture (particularly a close-up) as depicting only a part of a scene. As pointed out earlier, when we view a portrait, we perceive not only the details of the face, but that the rest of the person and the rest of the scene continue beyond the picture's boundaries. The schema is such an integral part of the perception of the picture, that subjects have difficulty in recalling or recognizing the picture's actual boundaries, and tend to incorporate the schema into their recollection. This moves the boundaries outward and results in boundary extension.

6. Implications for eye movements

The results of the boundary extension experiments suggest that a perceptual schema is important to picture perception and memory. If we grant the analogy between a picture and an eye fixation raised at the beginning of the chapter, the possibility that this schema may play a role in the integration of successive eye fixations can be considered. The first question that must be addressed in evaluating this possibility is whether the rapid, dynamic nature of visual scanning is compatible with the use of such a schema. After all, subjects can fixate a display up to 3-4 times per second. Yarbus (1967), for example, reported that subjects make approximately 3 eye fixations per second when studying a scene. It is necessary to consider whether there would be enough time available between the onset of one fixation and the onset of the next; a) to generate a schematic structure, and b) to make use of it to interrelate successive views. The picture perception literature provides some strong insights into these questions.

In one line of research, using tachistoscopic presentation of pictures, Biederman and his colleagues (e.g., Biederman, 1981; Biederman, Mezzanotte, & Rabinowitz, 1982) have demonstrated that the "gist" of a complex outline scene is perceived with remarkable rapidity. For example, Biederman et al. (1982) presented subjects with outline drawings of scenes for 150 msec each, followed by a mask and a visual cue. Subjects had to indicate whether the cued object corresponded to an object named at the trial's onset. Object identification took longer and was less accurate when the object appeared in an improbable context (e.g., fire hydrant in kitchen) than in a probable one (e.g., fire hydrant in street scene), even though it was presented in the same spatial location on the screen. They argued that gist acquisition had occurred rapidly enough to affect the viewer's speed and accuracy in identifying a specified object *within* this briefly glimpsed scene.

Based on similar experiments, Biederman (1981) has proposed that knowledge of real world scenes includes two general types of relationships among objects that can be referred to as "physical" and "semantic". He argued that in a meaningful scene, objects must follow physical laws of placement (e.g, they are supported, and they occlude the background). They also must follow semantic laws related to their identity (e.g., regarding their size or position with respect to other objects in the scene, and their probability of appearing in such a scene). He and his colleagues (e.g., Biederman, 1981; Biederman, Mezzanotte, & Rabinowitz, 1982) have studied the effects of violating these proposed physical and semantic expectations on the speed and accuracy of object identification. Using the same basic design described earlier, they demonstrated that violations of support, size, location, superposition, or probability all slowed or disrupted identification of a visually cued object in displays presented for only 150 msec. Violation of more than one of these laws at a time yielded even more interference with object identification.

This research strongly suggested that subjects' comprehension of briefly flashed outline scenes was not based upon identification of single objects, but that the physical and conceptual structure of the scene was grasped at a more general level. This relatively abstract "gist" could then guide perception of specific details within the scene. Biederman (1981) has likened this general scene knowledge to the concept of "deep structure" in language comprehension. Because the "gist" and the rules of scene structure affected object identification given such a briefly presented, masked stimulus, it seems plausible that these processes could take place for each fixation during visual scanning. Picture perception experiments using dynamic presentations of sequential stimuli, carry this point further.

Visual search experiments, using high-speed presentation of pictures, have demonstrated that unrelated scenes can be understood at presentation rates that surpass the average fixation frequency of the eye (Intraub, 1981a, 1981b; Potter, 1976). In Potter's (1976) research, color photographs of scenes were presented at rates ranging from 113 to 333 msec per picture (with no interstimulus interval). Subjects were presented with a number of sequences, each containing 16 unrelated scenes. A picture in the sequence was cued either by being presented in advance to the subject or by being described using a brief verbal title (e.g., "a road with cars"). The rationale for using both visual and verbal cues was that in order to match the verbal cue, the subject would have to understand the picture's meaning, whereas a match with the visual cue could be made on the basis of physical characteristics alone. Comparison of the two conditions was thought to provide insight into how much the search task relies on specific visual expectancies versus actual scene comprehension. The proportion of targets detected based on the verbal cue was interpreted as reflecting the minimal proportion of pictures identified during presentation.

To indicate detection, subjects pressed a key as soon as they saw the cued picture in the sequence. Responses falling between 250 and 900 msec following target onset were counted as correct. Results showed that subjects could identify unrelated scenes remarkably well at presentation rates that were equivalent to the average eye fixation frequency, and at those that were considerably more rapid. On the basis of the verbal cue, detection accuracy for rates of 113, 167, 250 and 333 msec/picture was 64%, 74%, 89%, and 78%, respectively (the apparent decrease at the slowest rate was due to an increased number of anticipation responses). Performance with the verbal cue was almost as good as seeing the picture itself in advance. These results showed that completely unrelated scenes presented at rates equal to or faster than the average fixation frequency of the eye could be conceptually analyzed and matched to a verbal cue. These results were replicated and extended by Intraub (1981b) using a number of design modifications, providing a further demonstration of the viewer's ability to rapidly grasp the meaning of a picture.

Intraub (1981b) used a "negative detection" task to determine if observers could identify unrelated photographs presented at high speeds, without receiving any direct information regarding their probable physical features. Although Potter (1976) had used general titles, one could argue that "a road with cars" narrows the visual expectations of the viewer and may enhance perception of the target picture. To minimize expectancy, in the negative detection experiment, a diverse set of 11 objects from a single general category and 1 picture (the target) that was not from that category were presented in rapid succession. Sequences were presented at rates of 114, 172, and 258 msec/picture. Photographs were selected that were as visually dissimilar as possible. For example, the category "animals" contained creatures as diverse as a frog, a dog, a giraffe, and a butterfly. The target picture did not differ in size or general color from the other pictures in the set. The cue provided at the beginning of each sequence was a "negative cue", for example, "the picture that is *not* an animal". Because no information about the target object was provided, correct detection responses were based upon the subject's description of the target, rather than inferring responses from reaction times. (Although reaction time was measured as well.) The results were consistent with those of Potter (1976).

Subjects were able to detect and describe target pictures surprisingly well at all three rates, without having been provided with any specific information about their visual characteristics. At the rate most closely approaching the average eye fixation frequency (258 msec/picture), 79% of all targets were detected with the negative cue. When the rate was increased to 172 and 114 msec per picture, far faster than observers can shift fixation during visual scanning, 58% and 35% of the targets were detected, respectively.

The search experiments show that a sizeable proportion of completely unrelated views can be rapidly understood when presented at rates of 3-4 pictures per second. Considering the fact that successive eye fixations on a real-world scene present the viewer with highly related, redundant, and overlapping views, the likelihood is that the ability to perceive the meaning and structure of each successive input is very high. Biederman's (1981) observations, taken in conjunction with the search experiments, paint a picture of rapid schematic analysis during briefly glimpsed pictorial presentations. This is consistent with Hochberg's (1978, 1986) proposal of a mental schema that guides and integrates successive views during motion picture viewing.

Returning to Hochberg's (1986) discussion of the use of a mental schema in aperture viewing, it is worthwhile to consider one other temporal aspect of picture processing. Recall that in the aperture demonstration, when subjects were informed that the ambiguous display they would see was actually an outline cross moving behind an aperture, they could only perceive the display as such when the presentation rate was 2-3 views per second. At more rapid rates of presentation the views became visually integrated and could not each be evaluated in terms of the mental schema (e.g., at 10 views per second). Using photographs and outline drawings of scenes like those used in the experiments just described, Intraub (1985, 1989) has reported perceptual integration of components of temporally adjacent displays at a similar presentation rate (9 pictures/sec).

For example, Intraub (1985) presented subjects with color photographs of objects at a rate of 9 picture/second. They were instructed to report which object in each 12-picture sequence was surrounded with a black outline frame. Although subjects were often confident and correct (50-70% of the time), they were frequently confident and wrong. Subjects, when wrong, almost always reported the immediately preceding or immediately following object in the sequence as the one with the frame. It was demonstrated that when subjects made an error, they often reported the target object (the one actually in the black frame) as a "frameless" picture. A similar effect was obtained when subjects were required to search for a specified object in a sequence of outline scenes (Intraub, 1989). Yet if these stimuli are presented at a rate of 3-4 pictures per second, the searched-for element (frame or object) does not become integrated with temporally adjacent pictures. At rates such as these, the successive views have apparently reached a state of analysis that protects them from such visual intrusions. It is at this rate, if we consider Hochberg's aperture experiment, that individually glimpsed views can be matched to an abstract schema depicting the layout of a scene -- a scene that is only partially depicted in each single view.

Experiments designed to test the effect of context on the pattern and duration of eye fixations, have also provided evidence for early gist acquisition and have demonstrated a tendency for observers to fixate longer on improbable objects in scenes, perhaps in part because of the longer processing time required to identify them (Antes & Penland, 1981; Friedman, 1979; Loftus & Mackworth, 1978). To conclude, it seems reasonable to argue that the timing limitations imposed by the dynamic nature of visual scanning do not rule out the use of a perceptual schema to guide and integrate successive fixations.

In future research, we will test the existence of boundary extension following brief stimulus presentations that are similar to a single eye fixation in duration. If boundary extension plays a role in the comprehension of successive views, then it should be detectable this early in processing. It is encouraging to note that in a recent replication of the research conducted with Berkowits, we found that decreasing the presentation duration of close-ups from 15 sec each to 4 sec each, had no effect on the magnitude of the boundary extension effect, as tested using the recognition procedure.

Drawing data are currently being analyzed. Presentation durations in the range of 100 to 500 msec per picture have yet to be tested. In other research, we are attempting to isolate conditions that will affect the degree or direction of boundary distortion for a given picture type. Once this is determined, the eye movement patterns for the different conditions will be compared, to directly study the inter-relationship (if any) of the schema and the actual locations fixated in the picture space.

7. Conclusions

Observers tend to remember a picture as having shown more of a scene than had actually been the case. They incorporate into their recollection, information that would have been likely to have existed in the scene just outside of the picture's boundaries. This phenomenon, referred to as "boundary extension" can be observed both in drawings and in recognition tests within minutes of picture viewing.

New research was reported that supports the hypothesis that boundary extension is, in part, the result of the activation of a scene schema during picture perception. A picture depicts only a part of a scene and contains this partial information within its boundaries. Represented in the schema are expectations about the form and structure of the scene that "exist" just outside those boundaries. It was proposed that this same schema may provide the basis for the perception of successive views of a scene during visual scanning. If the input of each fixation is considered to be a "picture" whose boundaries are determined by limitations of visual acuity, then the perceptual schema would function much as it does in the case of picture perception. It could thereby provide an abstract spatial context for the integration of successive fixations.

Further research into the nature of boundary extension may therefore provide a new means for studying the cognitive representation of scenes and its impact on visual perception.

References

Antes, J. R., & Penland, J. G. (1981). Picture context effects on eye movement patterns. In D. F. Fisher, R. A. Monty, & J. W. Senders (Eds.), *Eye movements: Cognition and visual perception*. Erlbaum: Hillsdale, NJ.

Bartlett, F. (1932). *Remembering: A study of experimental and social psychology*. England: Cambridge University Press.

Biederman, I. (1981). On the semantics of a glance at a scene. In M. Kubovy & J. R. Pomerantz (Eds.), *Perceptual organization*. Hillsdale, NJ: Erlbaum.

Biederman, I. (1987). Recognition-by-components: A theory of human image understanding. *Psychological Review, 94*, 115-147.

Biederman, I., Mezzanotte, R. J., & Rabinowitz, J. C. (1982). Scene perception: Detecting and judging objects undergoing relational violations. *Cognitive Psychology, 14*, 143-177.

Davidson, M. L., Fox, M. J., & Dick, A. O. (1973). Effects of eye movements on backward masking and perceived location. *Perception and Psychophysics, 14* 110-116.

Ellis, W. D. (Ed., Trans). (1955). *A source book of Gestalt psychology*. London: Routledge & Kegan Paul.

Friedman, A. (1979). Framing pictures: The role of knowledge in automatized encoding and memory for gist. *Journal of Experimental Psychology: General, 108*, 316-355.

Gibson, E. (1969). *Principles of perceptual learning and development*. New BF 311. G48 York, NY: Appleton.

Gibson, J. J. (1950). *The perception of the visual world*. Boston: Houghton Mifflin.

Haber, R. N. (1983). The impending demise of the icon: A critique of the concept of iconic storage in visual information processing. *The Behavioral and Brain Sciences, 6*, 1-54.

Hochberg, J. (1978). *Perception* (2nd ed.). Englewood Cliffs, NJ: Prentice-Hall.

Hochberg, J. (1986). Representation of motion and space in video and cinematic displays. In K. J. Boff, L. Kaufman, & J. P. Thomas (Eds.), *Handbook of perception and human performance* (Vol. 1, pp. 22:1-22:64). New York: John Wiley & Sons.

Intraub, H. (1981a). Identification and processing of briefly glimpsed visual scenes. In D. F. Fisher, R. A. Monty, & J. W. Senders (Eds.), *Eye movements: Cognition and visual perception*. Erlbaum: Hillsdale, NJ.

Intraub, H. (1981b). Rapid conceptual identification of sequentially presented pictures. *Jounral of Experimental Psychology: Human Perception and Performance,* 7, 604-610.

Intraub, H. (1985). Visual dissociation: An illusory conjunction of pictures and forms. *Journal of Experimental Psychology: Human Perception and Performance, 11*, 431-442.

Intraub, H. (1989). Illusory conjunctions of forms, objects and scenes during rapid serial visual search. *Journal of Experimental Psychology: Learning, Memory, and Cognition, 15*,98-109.

Intraub, H., Bender, R., & Mangels, J. (1992). Looking at pictures but remembering scenes. Journal of Experimental Psychology: Learning Memory and Cognition, *18*, 180-191.

Intraub, H., & Richardson, M. (1989). Wide-angle memories of close-up scenes. *Journal of Experimental Psychology: Learning, Memory, and Cognition, 15*(2), 179-187.

Irwin, D. E., Brown, J. S., & Sun, J. (1988). Visual masking and visual integration across saccadic eye movements. *Journal of Experimental Psychology: General, 117*, 387.

Irwin, D. E., Yantis, S., & Jonides, J. (1983). Evidence against visual integration across saccadic eye movements. *Perception and Psychophysics, 34*, 49-57.

Loftus, G. R., & Mackworth, N. H. (1978). Cognitive determinants of fixation location during picture viewing. *Journal of Experimental Psychology: Human Perception and Performance, 4*, 565-572.

Neisser, U. (1967). *Cognitive psychology*. Appleton-Century-Crofts: New York.

Potter, M. C. (1976). Short-term conceptual memory for pictures. *Journal of Experimental Psychology: Human Learning and Memory, 2*, 509-522.

Shepard, R. N. (1967). Recognition memory for words, sentences, and pictures. *Journal of Verbal Learning and Verbal Behavior, 6*, 156-163.

Standing, L., Conezio, J., & Haber, R. N. (1970). Perception and memory for pictures: Single trial learning of 2,560 visual stimuli. *Psychonomic Science, 19*, 73-74.

Yarbus, A. L. (1967). *Eye Movements and Vision*. New York: Plenum Press.

The Role of Eye Movements in Perceptual Processes
E. Chekaluk and K.R. Llewellyn (Editors)

EYE MOVEMENTS AND PERCEPTUAL MULTISTABILITY

MIGUEL A. GARCIA-PEREZ

1. Introduction

A multistable or reversible pattern is one that can give rise to more than one interpretation. Natural scenes of do not contain any such patterns. The real objects and creatures that surround us in natural settings are so rich in details that they hardly admit other interpretations than the veridical. Our visual system provides us with either that interpretation or none at all, in case we look at something from an unconventional viewpoint or we face something of a class we have never seen before. In contrast, for the most part, the phenomenon of perceptual multistability occurs with a number of articifically created patterns that are rough sketches of their real-world referents. It may be argued that the association of multistability with such artificially created patterns implies that the phenomenon tells us little about the normal operation of the visual system. Yet, the fact that it does occur, even if under artificial conditions, can provide useful information.

It is important to note that there are two different problems which require explanation about perceptual multistability. The first one is how it is that we can interpret a rough sketch as something whose real-world referent is much richer in details. This interpretational problem does not arise only when viewing multistable figures, but results when inspecting any line drawing.

That is why this first problem will not be dealt with in this chapter. The second problem is unique to multistable figures, and it arises from the fact that *the same* pattern can be interpreted in at least two different ways, only one of which is available at a given point in time. The most striking fact about this multiplicity of interpretations is that prolonged inspection of the figure results in an alternation of the competing interpretations. The second and main question is, then, what is it that causes the alternation, and it is research on this problem that will be discussed here.

Multistable figures set a challenge both to the visual system and to the visual scientist. The visual system has to deal with and interpret them, and the visual scientist must explain how it is that the visual system does so.

This paper was prepared under a grant from Comision Interministerial de Ciencia y Tecnologia, Ministerio de Educacion y Ciencia.

Although the visual system handles its part of the challenge rather expeditiously, visual scientists are taking much longer with theirs. Visual illusions present a similar problem. In their review of the state of the art, Coren and Girgus (1978) discussed evidence suggesting that every stage of visual processing contributes to the visual illusions. They pointed out that, by combining different research methodologies, the percentage of contribution to the illusory effect that is due to optical, neural, and cognitive-judgmental factors can be estimated. The influence of cognitive factors was, however, assessed by eliminating the measurable contribution of optical neural factors. It therefore remains to be unravelled what the so-called 'cognitive factors' are and how they act.

Research on perceptual multistability has largely focused on the cognitive or top-down processes that provoke the phenomenon. By contrast to the emphasis on the role of optical and neural factors in the study of visual illusions, research on the role of optical and neural factors in perceptual multistability has received very little attention thus far.

This chapter attempts to summarize our knowledge on multistable perception of stationary patterns, with special emphasis on the role of eye movements in solving multistable figures. A classification of multistable figures is presented first. Then, the literature on perceptual multistability is reviewed, summarizing the known facts about this phenomenon and the variables that have been shown to affect it. Next, the characteristics of early spatial visual processing are presented along with a model that describes quantitatively the perceived image of a scene. Finally, and given the perceived image of a multistable figure, the presumed role of eye movements in solving them is discussed.

2. Types of Multistable Figures

Perhaps the only feature that multistable figures share is their defining property of being capable of admitting several interpretations and actually giving rise to their alternation as a result of prolonged inspection. Yet, it is feasible to attempt a classification of multistable figures in terms of the property of the percept that changes. Nonetheless, it is important to keep in mind that the resulting classification does not necessarily reflect equivalence classes such that figures in a class are dealt with by the visual system in a fixed way that differs from the way that figures in other classes are treated. This is, then, a textbook classification of multistable figures rather than one based on their visual processing. It gives rise to the five categories that are described next.

Some moving patterns are subject to *stereokinesis*. Braunstein and Andersen (1984) showed that a rigid two-dimensional shape that is rotated in the frontal plane is perceived as a distorting three-dimensional shape. Some of these stereokinetic effects show multistability. Bressan and Vallortigara (1986) noted that when a solid flat ellipse is rotated in the frontal plane, it can be perceived either as an elongated egg slanted in three-dimensional space and describing a circular trajectory in the frontoparallel plane or as a circular disc oscillating back and forth in three-dimensional space. Because it is the analysis of motion that makes the interpretation change, an explanation for this type of multistability has to be found in the visual-motion processing system. Since this chapter will only consider the spatial processing system, that type of multistable pattern is the only one among the five types presented here that will not be dealt with in the remainder of the chapter.

Some stationary patterns admit *figure-ground reversibility*. Patterns giving rise to this phenomenon consist of large black and white areas whose boundaries are drawn in a way that seeing a white figure on a black background is as possible as seeing a black figure on a white background. The Rubin vase/face and a number of drawings by M.C. Escher are well-known examples of this type of multistable figure. The effect also occurs if outline figures on a white background are used, that is, there are no areas of black.

Various geometrical shapes elicit *orientation reversals*. This multistability can be cast in terms of the assignment of reference frames for the description of the shapes involved. For example, the assignment of up, down, left, and right directions determines whether Mach quadrilaterals will be seen as diamonds or squares. Also, an equilateral triangle can be seen as an arrowhead pointing in any of the three directions signaled by its vertices.

Some two-dimensional sketches of three-dimensional objects are *perspective-reversible figures*. In these drawings, recovery of the three- dimensional structure from the two-dimensional rendering can be accomplished with the same effort in at least two ways. The Necker cube and the Schroder staircase are good examples of this capability. An interesting fact about Necker cube (which may also apply to some other persertive-reversible figures) is that when the figure rotates, it changes perceived direction of rotation whenever perspective changes (Howard, 1961). Because the visual-motion processing system is also involved in this phenomenon, research on it (e.g., Long, Toppino & Kostenbauder, 1983) will be left aside here.

The last type of multistable stationary pattern is often called *ambiguous figure*. Quite often, these figures are sketchy drawings consisting of small line segments, either connected or not. Separately considered, each line segment has no meaning at all, but when they all are taken together, every single line admits more than one interpretation as a part of a larger drawing representing a sketch of a real-world percept. All lines play one of their multiple roles when one of the interpretations is available, but they all change their meaning at once when the figure reverses. Some of these ambiguous figures are the Boring old/young woman, the Fisher man/girl, or the duck/rabbit figure. Unlike the remaining multistable stationary patterns, ambiguous figures also exist as gray-level or full-color pictures. In this case, it is the texture fields, edges and shadows that can be put together in several ways to yield two or more completely different percepts. Salvador Dali's *Slave Market with Apparition of the Invisible Bust of Voltaire* is a famous example of this type of picture.

3. Perceptual Multistability: Phenomena and Explanations

In doing research on reversals in perceptual multistability, it is important to define two measurable quantities that describe the phenomenon. One of them is the *reversal rate*, that is, the number of reversals that occur per unit time. It can be defined as the inverse of the time interval separating three consecutive reversals. These three consecutive reversals define two time intervals, each corresponding to the perception of a different organization of the multistable figure. An experimental measure related to this reversal rate is the overall number of reversals that occur after inspection of the figure for a given time period, although reversal rate may change during that period.

A second quantity is *percept duration*. Percept duration is defined as the amount of time between two consecutive reversals, that is, the amount of time that a given organization of the figure lasts before it is replaced with its competitor. Percept duration is not necessarily the same for all possible percepts associated with a multistable figure. An experimental variable related to percept duration is the overall percept duration, or overall time during which a given interpretation of the figure occurs in the inspection period, although percept duration is also subject to changes throughout that period. Recently, Shank and Walker (1989) defined some new measures of the strength and ambiguity of multistable figures based on these quantities.

Other than Attneave's (1971) and Vicker's (1972) papers on the subject, a review of research on perceptual multistability has not been published as far as the present author knows. An attempt in that direction is made here. The review is organized in a way consistent with the goal of the paper, namely, discussing the role of eye movements in solving these figures. As such, the review is not complete. The author apologizes to colleagues whose preferred paper is not discussed here.

3.1 Dominance of aspect

One thing in which multistable figures differ from one another is in what has been called 'dominance of aspect'. This refers to the imbalance between percept durations for each possible interpretation, and also to the imbalance between the probabilities of seeing one or the other interpretation at first encounter. Fisher (1967) measured this dominance of aspect for his man/girl figure, and devised a biasing technique that affects it.

Generally speaking, the type of multistable pattern that is most subject to dominance of aspect is the perspective-reversible figure, and the interpretation that is more dominant is that of a three-dimensional object seen from above. This may be because that is the way we normally encounter the implied three-dimensional objects in the real world. Similarly, orientation-reversible shapes are usually interpreted according to an implicit gravitational vertical. For figure-ground reversible figures, the black area is generally more likely to be seen as figure. Finally, ambiguous figures are not subject to a strong dominance of aspect.

3.2 Rate of spontaneous reversals

Because of this dominance of aspect, some multistable figures do not in fact appear to be multistable to the naive viewer (Girgus, Rock & Egatz, 1977). Quite often, subjects arrive at one of the interpretations immediately upon seeing the figure for the first time, and they stick to that interpretation until some hints are provided to help them interpret the figure the other way. The alternation only occurs when the subject is aware of the different interpretations that are possible, and then it occurs spontaneously.

Using prespective-reversible figures (a Necker cube, a Schroder staircase, and a honey-comb figure) Borsellino, De Marco, Allazetta, Rinesi and Bartolini (1972) measured the rate of these spontaneous reversals and the overall percept duration for each interpretation. They found that the rate of alternation increased rapidly during the first two or three minutes of observation, reached a stationary state, and then decreased slowly.

Subsequent analyses (De Marco, Penengo, Trabucco, Borsellino, Carlini, Riani & Tuccio, 1977) showed that gamma functions were capable of describing the distribution of time intervals between consecutive reversals in the time band where multistability is in a stationary state.

One problem in interpreting these results in that it is not clear how spontaneous those reversals actually were. Although subjects were intructed to maintain fixation and refrain from exerting voluntary control over the alternation throughout the lengthy sessions, they still might have caused some of them unwillingly. Knowing that one is expected to report changes in the interpretation of the figure puts a little strain on oneself while waiting for the change to occur, and just thinking of the alternative interpretation might bring it forth. Nonetheless, conducting an experiment like this under fully controlled conditions does not seem feasible.

3.3 Voluntary control over reversals

As noted above, alternation does not occur only in spontaneous fashion. A number of studies have addressed the question of the extent to which reversal rate and percept duration can be controlled at will. There are a number of instruction sets that can be given to experimental subjects to make them develop and use cognitive strategies that will affect the rate of alternation and the duration of each percept as compared to the same variables when subjects observe passively.

Pelton and Solley (1968), instructed subjects to change perspective of a Necker cube as often as possible. These subjects reported more reversals than did other subjects instructed to maintain their experienced perspective, whichever it was at each point in time, as long as possible. Also, using instructions to the effect of maintaining a given percept of the stimulus and trying to switch back to it as soon as a reversal occurred, Peterson and Hochberg (1983) found that the duration of the aimed perspective of a Necker cube could be made longer than it is when no such instructions are given.

Phillipson and Harris (1984) conducted a similar study on the Schroder staircase. In one experiment, they asked different groups of subjects to try to speed up reversal rate, to try to slow it down, or to remain passive. Results showed that the mean number of reversals can be increased or decreased by about 30% with the help of this mental effort. At the same time, they found that these changes in reversal rate did not affect the percentage of viewing time during which each of the percepts was seen. In another experiment, where the subjects were asked to maintain one of the percepts as long as possible and change back to it whenever a reversal occurred, they found changes in percept durations that were consistent with the instructions given. In addition, compliance with the strategy induced by these instructions made mean reversal rate decrease by about 30%. Altogether, the results of these two experiments show that both reversal rate and percept duration can, within limits, be controlled at will. In addition, voluntarily induced changes in reversal rate do not result in changes in overall percept duration, while voluntary control over this duration does affect (lower) reversal rate. Liebert and Burk (1985) found a similar increase of 20% in the mean duration of the Schroder staircase. They also found an increase of 66% with the reversible screen, which could reflect the fact that the instructions given (trying to bring a feature of each figure closer in perspective) were more appropriate for the screen than they were for the stair. It was also found in this study that the mean number of reversals did not change as a function of instructions, and that individual differences exist as to the capability of exerting voluntary control over reversals.

Altogether, these results show that voluntary control over reversals is possible, although it does not override completely other sources of alternation. The main problem posed by these results is the determination of the processes that underlie this voluntary control. The only thing we have so far been able to say about it is that voluntary control occurs because there is something that accomplishes it, although we are prone to call that something a 'cognitive set' or a 'mental effort' to hide the inherently *ad hoc* and circular nature of this "explanation". A true explanation would require actual (as opposed to metaphorical) specifications as to what processes are working, and in what way, to accomplish the control, both of the reversal rate and of the time spent on each interpretation. Any such explanation should also account for the failure of those processes to fully determine perception when they are brought into play.

3.4 Biochemical bases of perceptual multistability

As is the case for all psychological phenomena, perceptual multistability must have some correlate in neural activity in the brain. If the alternation is caused by competition between or within neural networks, then its dynamics should be affected by controlling the availability of the neurotransmitters on which those networks operate.

Inspired by the unusual way in which psychotic patients process ambiguous material (mostly homonyms), Phillipson and Harris (1984) measured the effect of chlorpromazine (a dopamine receptor blocker) and promazine (a central adrenergic receptor blocker) on the perception of the Schroder staircase (a perspective-reversible figure) in healthy subjects. They found that both drugs not only lowered reversal rate but also increased dominance of aspect. Similar results were found for the Fisher man/girl, an ambiguous figure. In summary, these results indicate that the drugs accentuate an existing perceptual bias or proneness towards one of the interpretations, and reduce the rate of alternation. The effects just described are not strong enough to claim that the still unspecified neural networks involved and their dynamics are the only cause of the alternation, but they definitely show that neural activity has something to do with them, and they encourage the study of neural networks capable of multistable behaviour. Yet, as Phillipson and Harris (1984) pointed out, it is not clear from these results whether the drugs affect the mechanism responsible for the spontaneous alternation, the mechanism involved in its voluntary control, or the balance between both mechanisms.

3.5 Variations on multistable figures

Many variations on multistable figures have been used in experimental studies to assess the influence of figural changes on perceptual multistability. For the most part, these changes introduce a kind of bias in the stimulus that makes one of the interpretations more likely then the others. That is, these variations change the dominance of aspect as compared to that of the unbiased multistable figure. In principle, introduction of these changes was meant to check the extent to which the different features in the multistable figure elicit its different interpretations. Results of this research, however, are not as illuminating as one would like them to be.

Klymenko and Weisstein (1986) prepared versions of figure-ground reversible figures in which the black and white areas were replaced with sinusoidal gratings of different spatial frequencies. They found that this filling-in served as a strong biasing factor in that the region which had the higher spatial frequency turned out to be strongly perceived as figure.

In the case of orientation-reversible figures, there are many different ways in which bias can be introduced. One way is to arrange several figures of the same shape in certain configurations. Attneave (1968) showed that groups of equilateral triangles randomly located are subject to the same multistability as a single triangle. Palmer (1980; see also Palmer & Bucher, 1981) found that linear configurations of triangles aligned by one of their sides or axes can bias perceived orientation strongly. Similar configural biases were described for Mach quadrilaterals (Palmer, 1985). An important finding of this work is that the biasing influence of configural information is asymptotical in that the effect produced by the arrangement of a certain number of elements cannot be increased by adding more elements to the configuration. The asymptote is reached when the number of elements is such that the whole configuration extends to two degrees of visual angle away from the fixation location.

One other way to introduce bias in the perception of multistable geometrical shapes is to add some textural context to the drawing of a single shape. Palmer and Bucher (1982) and Palmer (1985) added patches of square-wave gratings at different orientations either inside or surrounding the figure (equilateral triangles and Mach quadrilaterals, respectively) and found similar effects as for configural context. They also found the asymptotic effect mentioned above.

Finally, bias can be introduced by adding some contextual information that allows the figure to be seen in a three-dimensional environment (Palmer, Simone & Kube, 1988). These three-dimensional cues exert the expected influence on the dominance of one of the interpretations of the ambiguous figures.

Bias has been added to perspective-reversible figures in two ways. One way (Harris, 1979) is to draw the figure as if it was seen from a point of view different from the conventional (unbiased) one in which the figure was originally created. This results in a measurable change of perspective whose effect can be assessed experimentally. This bias was also manipulated by adding some textural gradients appropriate to the depth cue given by the changed perspective. Both the number of reversals and the dominance of aspect of a Schroder staircase were shown by Harris (1979) to change consistently and monotonically with the amount of perspective bias introduced in the figure.

The other way is to introduce in the figure contextual cues that will affect the likelihood of each three-dimensional interpretation. Riani, Tuccio, Borsellino, Radilova and Radil (1986) reduced the ambiguity of a Mach truncated pyramid and a modified Necker cube in several degrees either by adding new elements to the figure that would favor one of the interpretations or by removing items from it until a percept different from either interpretation of the ambiguous figure was reported. Measures of reversal rates and percept durations in each case showed significant effects of the biasing elements: decreased reversal rate and increased duration of percepts. Subsequent work (Masulli & Riani, 1989) allowed these effects to be related to the complexity of the figures as measured through the coding system provided by structural information theory (Buffart & Leeuwenberg, 1983).

Ambiguous figures can be biased towards one or the other interpretation using the technique developed by Fisher (1967). It consists of progressively modifying the different features that make up the figure in a way that it suggests one interpretation or the other. In the end, a whole series of figures can be created that range from a 100% unambiguous version eliciting one of the interpretations of the multistable figure to a 100% unambiguous version eliciting the other interpretation through a middle point at the original, purely ambiguous figure.

The conclusion of all of these studies is that all forms of bias disambiguate any ambiguous figure. Bias can be added to different degrees, and there seems to be an asymptote in bias that functionally corresponds to the loss of the multistable nature of the figures: any further bias introduced in the figure cannot increase an already maximum effect.

3.6 Stimulus variables that affect reversals

The role of stimulus variables that are known to affect the pattern of neural excitation within the visual system has also been studied. Any observed change in the rate of alternation or the dominance of aspect that is associated with these stimulus changes might then indicate the type of visual processing that is related to perceptual multistability.

Lindauer and Lindauer (1970) measured the influence of the contrast of a Rubin vase/face figure on reversals. Contrast was manipulated by changing the gray level of the dark area. They found that, on first encounter, the dark area tended to be seen as figure and that this preference to see the dark area as a figure increased with increasing contrast. However, no effect of contrast was found on percept duration or number of reversals. Shank and Walker (1989) confirmed the influence of contrast for a new figure-ground reversible figure that they had created.

Because of the nonuniform distribution of rods and cones in the retina and of the non-linear mapping of the retina onto the visual cortex, the cortical representation of a stimulus changes with its retinal size. Borsellino, Carlini, Riani, Tuccio, De Marco, Penengo and Trabucco (1982) measured the effect of the size of a Necker cube on the pattern of reversals. Their results show that the duration of each percept does not change with figure size; however, reversal rate, which was about constant for all sizes up to 20-30 degrees of visual angle, decreased with further increases in size.

The involvement of rod and cone vision was further investigated by Riani, Oliva, Selis, Ciurlo and Rossi (1984) using a Necker cube. Luminance ranged from values appropriate for photopic vision (when cones are activated) to values for scotopic vision (when cones are not stimulated). Results showed no effect of mean luminance either on reversal rate or on percept duration. Therefore, the processes responsible for the alternation are not dependent on whether it is the cones or the rods in the retina that encode the figure.

Visual processing is currently understood in terms of spatial frequency processing as performed by a bank of channels, each of which is selectively sensitive to a narrow range of spatial frequencies and orientations (see Kelly & Burbeck, 1984). Therefore, one other way in which the role of visual processes in perceptual multistability can be assessed is by changing the spatial frequency content of the figure so that some of these channels are not activated. Petersik (1979) recorded the pattern of reversals elicited by a dynamic process of defocusing and refocusing of a perspective-reversible figure akin to a Necker cube. This manipulation makes the visual system go through a sequence of stages in which the higher spatial frequency channels are gradually deprived of input and provided back with it. The defocusing and refocusing was done while subjects were voluntarily maintaining one of the interpretations of the figure. Results showed that at some point during defocusing, which turned out to be the same for all subjects, an automatic reversal occurred. Also, during refocusing, a constant point was also found for all subjects in which an automatic reversal took place. These results point to the fact that high spatial frequency (above 1.15 cycles per degree) channels (or the balance of their activity with that of low spatial frequency channels) are involved in reversals.

3.7 Procedural variables that affect reversals

Research has also shown that the experimental conditions under which the task of inspecting multistable figures is studied affect the measurable quantities describing the phenomenon.

Reisberg and O'Shaughnessy (1984) measured the multistability of perspectve-reversible figures when subjects were engaged in an arithmetic distractor task. Results showed that reversal rate was slowed significantly by the distractor task, as if a reversal demanded attention that was not available because of its allocation to the arithmetic task.

Epstein and Rock (1960) presented an alternating sequence of unambiguous versions of ambiguous figures before the purely ambiguous figure was presented, and showed that there is a recency effect whereby subjects perceived the ambiguous figure as an instance of the unambiguous version experienced most recently. These results allowed them to reject an effect of expectancy, which would have predicted that subjects should have seen the figure as an instance of the unambiguous version that should have appeared in the sequence. Work by Wilton (1985) showed that whether a recency or an expectancy effect is found depends on variables such as the duration of the unambiguous stimulus immediately preceding the ambiguous figure, or the time interval between the next-to-most-recent unambiguous stimulus and the ambiguous figure. The duration of the ambiguous figure itself has no influence on the recency/expectancy effect expect when the stimulus is exposed several times in the experiment.

Research involving biased and unbiased versions of multistable figures has also shown a priming effect in the interpretation of a figure at first encounter. Owen (1985) presented biased versions of ambiguous figures (Boring young/old woman and the mouse/boy figure) and unbiased versions of them in close temporal contiguity. The biased versions were presented for a very short time to prevent their conscious processing. They were followed by a pattern mask and then the corresponding unbiased version was displayed for a longer time. Results showed that the interpretation reported for the unbiased figure was congruent with the biased figure that was presented as prime, even though the short presentation and the masking of the prime prevented it from entering awareness. The dichoptic presentation of the figures also suggests that this priming effect is operating at a stage of processing beyond the point where information from the two eyes is combined. Similar priming effects were found by Goolkasian (1987).

Stewart and Peregoy (1983) reported an interesting phenomenon of hysteresis. They used the series of drawings created by Fisher (1967) and asked subjects to rate each of the drawings in the series as a man or a girl. The rating scale was from 0 (ambiguous figure) to 10 (completely unambiguous figure, qualified as either a man or a girl). The sequence of drawings was arranged so that it started with the unambiguous drawing of a man's face, proceeded by increasing ambiguity until a purely ambiguous version was obtained, and then increased bias towards the other interpretation up to a purely unambiguous drawing of a kneeling girl. Subjects viewed either this sequence or one in the reverse order. Results showed that subjects always changed their perception beyond the middle point in the sequence, regardless of the order of presentation of the drawings.

3.8 Eye movements while viewing multistable figures

Hypotheses about the role of eye movements in solving multistable figures can only come from direct observation of eye movement patterns made while such figures are viewed. The earlier studies (Sisson, 1935; Pheiffer, Eure & Hamilton, 1956) were hindered by the unavailability of reliable eye-trackers, but still these studies allowed some preliminary data to be found which suggested that eye movements are somehow associated with reversals. The evidence provided by these studies was, however, equivocal as to whether eye movements are a cause or a consequence of reversals.

More recently, Ellis and Stark (1978) recorded eye scanpaths during the viewing of Necker cubes. A record was also kept of the instants at which reversals occurred, which allowed the spatial location and duration of fixations at the instant of a reversal to be determined. The results of their analysis are quite clear in two respects. Firstly, fixation duration is markedly longer when it immediately follows a reversal than it is when no reversal occurs, as if subjects needed some time to build up a replacement for the current interpretation. Secondly, fixation locations at the instant of a reversal are different for the two three-dimensional interpretations of the cube, as if looking at different parts of the figure selectively favored one interpretation over the other.

Sakano (1963) had already reported this latter fact: subjects asked to hold one of the interpretations of an ambiguous figure tended to fixate on specific parts of the stimulus. Also, Kawabata, Yamagami and Noaki (1978), on analyzing fixation locations while observing a Necker cube and a Schroder staircase, found that fixations on different parts of the figures tended to elicit different three-dimensional interpretations. Gale and Findlay (1983) also found evidence suggesting that fixation on different parts of an ambiguous figure (Boring young/old woman) helps hold each of its different interpretations.

Gale and Findlay (1983) obtained the currently most extensive set of results about eye movement patterns during the viewing of an ambiguous figure. On the basis of the evidence they obtained, the following conclusions can be drawn:

1) When instructed to concentrate on one or the other interpretation of the multistable figure, subjects' fixations are limited to distinctly different areas of the stimulus.

2) When subjects inspect the stimulus freely, the reported interpretation when fixations occur around a certain area is the same as that interpretation subjects tried to hold by looking at that particular area.

3) Eye movements resulting in a change of fixation from an area favoring one interpretation to an area associated with the other resulted in reversals. But reversals also occurred without being preceded by an significant change in fixation location.

These areas which selectively favor one interpretation of the figure, which are fixated when that interpretation is attempted to be held, and which favor a reversal when eye movements place fixation location on them, can be regarded as *focal areas* for the corresponding interpretation.

3.9 Locus of spatial attention and multistable figure perception

Some indirect evidence of the role of fixation in the perception of multistable figures comes also from studies in which the allocation of spatial attention on a multistable figure is controlled by the experimental procedure, without eye movements or actual fixation locations being recorded.

Tsal and Kolbet (1985) asked subjects to maintain each of the two interpretations of two ambiguous figures in separate blocks of trials. In each trial, an ambiguous figure was tachistoscopically replaced with a blank stimulus or with a letter either at the focal area of the instructed interpretation or at that of the alternative interpretation. Letters presented on the focal area of the instructed interpretation were significantly faster detected than those falling on the focal area of the alternative interpretation. In another experiment, Tsal and Kolbet (1985) reversed the order in which the letter and the ambiguous figure were presented in each trial. Subjects were asked to report the perceived interpretation of an ambiguous figure that was tachistoscopically presented after a letter was displayed. Display of the letter was meant to call subjects' attention to a spatial location on which the focal area for one or the other interpretation of the ambiguous figure would afterwards appear. Results showed that the most frequently perceived interpretation was that for which the letter signaled its focal area.

Kawabata (1986) further investigated the relationships between attention to focal areas and perceived interpretation with Necker cubes. He had subjects attend to local angles of briefly presented Necker cubes. The angle to be attended was signaled by drawing that particular angle with either thicker or thinner line segments, and was always presented at the fixation point. Results showed that attention to thick angles at the focal areas described by Ellis and Stark (1978) and Kawabata, Yamagami and Noaki (1978) gave rise to the expected interpretation. When thin angles were used, however, no facilitation was found for the perception of the expected interpretation, possibly because those thin angles were not very conspicuous.

Goolkasian (1987) had subjects direct attention to focal or neutral areas of ambiguous figures with the help of a fixation point. She found that the interpretation of the figure was faster achieved when the fixation point signaled a focal area. In another experiment, presentation of the figures was preceded by presentation of the elements comprising the focal areas, but only slight priming effects were found.

All in all, these experiments suggest that allocation of attention to specific parts of a multistable figure is related to the perceived interpretation of the figure, although it is not clear whether actual fixations are needed to accomplish this focusing of attention.

3.10 Proposed explanations for the alternation

Despite the overwhelming amount of data gathered on perceptual multistability, a more or less complete and realistic explanation for the phenomenon has not as yet been achieved. There are a number of attempts in that direction that nevertheless fail to provide a true explanation in terms of visual processes or explicit cognitive processes acting upon visual representations. The available explanations are either discursive or metaphorical.

Perhaps the metaphor that comes in handiest is that proposed by Kohler and Wallach (1944). They attributed the alternation to competition between structures that are subject to satiation and recovery. At first encounter with the figure, the structure responsible for seeing one of the percepts happens to be dominant and determines perception. While dominating, this structure gradually satiates and loses strength up to a point in which its strength becomes less than that of the competing structure, which takes over perception and causes a reversal. From that point on, the dominating structure begins to satiate while the other recovers and at some point in time the balance of their strengths changes again and a new reversal occurs. This process goes on throughout the inspection period.

Several authors have devised physical or neural models capable of behaving in a way congruent with this metaphor. This first physical model is probably the multivibrator circuit sketched by Attneave (1971), which is an electronic model that merely shows that multistable behavior can have a purely automatic nature. However, as a model of multistable perception, it lacks all psychological appeal.

Kawamoto and Anderson (1985) designed a neural network model governed by Hebbian-like synaptic modification rules which was shown to reproduce some experimental data on perceptual alternation. Although the neurophysiological metaphor is somewhat more appealing than the electronic one, and the mathematics of the neural network model gives it a strong outlook, this neural model is not very appealing psychologically because its input is not the visual representation of a multistable figure but a vector plucked out of the air and filled in with numbers conveniently chosen to make the network produce the desired output. Kienke, Sejnowski, Hinton and Schumacher (1986) designed a more down-to-earth parallel network acting on the stimulus proper that was capable of achieving figure-ground segregation. Their network was not capable of multistability, but they mentioned that an extension to it in line with Kawamoto and Anderson's (1985) ideas might provide it with multistable behavior. If it turned out to be true, then the resulting model would probably be the more complete one as an explanation of some of the facts about data-driven figure-ground reversals.

A somewhat different approach to modelling perceptual multistability is that of Vickers (1972), who proposed an information-processing model based on a cyclic decision process. The decision is taken on the difference between the values of random variables akin to those representing signal and noise in signal detection theory. These variables represent, in a metaphorical way, evidence in favor of one or the other interpretation of the multistable figure. As soon as a decision is made, the corresponding perpect is seen, and the whole process starts anew. A reversal occurs whenever the decision just made is different from the immidiately preceding one. The model includes several parameters that adjust its behavior in order to accommodate experimental data on reversals. This model, however, shares all of the flaws usually encountered in models proposed within the information processing paradigm (see Kolers & Smythe, 1984).

The models mentioned so far incorporate processes whose output variables have values describing the same pattern of changes as is found in a certain subset of experimental data on perceptual alternation, but they are obtained using a mathematical apparatus that does not represent actual visual or cognitive processes. Also, the models make no mention of how the input variables they are fed with are obtained from the multistable figures.

There is another type of model, even more simplistic, in which not a single process is mentioned, being limited to describing mathematically the patterns found in experimental data. This category includes Taylor and Aldridge's (1974) or De Marco, Penengo, Trabucco, Borsellino, Carlini, Riani and Tuccio's (1972) stochastic models of reversal time and Poston and Stewart's (1978) catastrophe theory model of multistability.

On considering these mathematical models, it should be kept in mind that "models are not necessarily more veridical or more to be believed by virtue of being mathematical" (Coombs, 1983, p.94), and that "not everything published in psychology with equations or mathematical terms in it is, necessarily, a serious or satisfactory attempt to involve mathematics in theory development". (Luce, 1985, p.669.) In fact, the models mentioned thus far are formulated mainly with a view to describing the facts and being able to mimic the experimental data, but they do not do so by representing actual and well-established visual or cognitive processes. Therefore, the models account for the general features of perceptual alternation, but they cannot be regarded as models of *multistable figure processing*. For this reason, these models cannot explain mechanistically how it is that the variables mentioned earlier in this section affect multistable perception. Those variables simply fall outside the domain of these models. Or, perhaps, it is more fitting to say that it is those models that fall outside the domain of reality.

The currently most widely accepted explanation for the alternation in perceptual multistability is of a discursive nature. It relies on the empirical fact that different areas of the stimulus favor different interpretations, and states that reversals are due to *shifts of attention* from one focal area to the other. The mechanism or set of mechanisms capable of accomplishing these attentional shifts remain largely unspecified, but a change in fixation location is surely one of them. In fact, as was discussed in Section 3.8, a change in fixation location from a focal area to the other does produce a reversal. The fact that reversals also occur without eye movements implies that other mechanisms must control the necessary attentional shifts too and, therefore, eye movements are not necessary conditions for reversals. Yet, as one of the parts of the attention process, it appears that they are sufficient conditions for the alternation.

Because eye movements are not the only instrument to accomplish an attentional shift, a less than perfect correlation between them and reversals has to be expected. In any case, the particular way in which a change in fixation location brings a reversal has to be specified. This is precisely what this chapter deals with by analyzing the visual information available at a glance. This analysis sets out the basis for complying with Peterson and Hochberg's (1983, p.192) claim that "coding theories of figure perception (...) must be redesigned to consider the limits on the information present within each glance."

4. Early Spatial Visual Processing

As was said in the introduction, the effect of the early filtering stages of visual processing on perceptual multistability has not been analyzed in depth. This section sets out the basis for such an analysis.

Early spatial visual processing deals with the building of spatial visual representations. To the eyes of a psychologically oriented visual sicentist- who does not care much about the internal format of representations- a spatial visual representation is the image that we actually perceive when we fixate on a particular location in a scene. That is what the world before our eyes looks like to us at a given point in time. Roughly speaking, the perceived image is sharp and clean around the fixation location, but it gradually blurs and loses detail with increasing distance to that location. This interest in the perceived image of a scene arises because it is that image (as opposed to the physical image that the visual system processes to produce it) that gives rise to the phenomena that perceptual psychologists want to explain. What we demand of a model of early spatial visual processing is that it describes quantitatively the luminance distribution in the perceived image of a scene given a fixation location.

Most of the spatial vision models in the literature have arisen from experiments involving grating-like stimuli at threshold contrast (see Kelly and Burbeck, 1984, p.126, for a recent listing of that kind of model). Although those models are capable of describing rather accurately the experimental data they were set up to account for, they nonetheless have a very restricted domain and they fail to provide a comprehensive view of visual processing at normal contrast conditions. In addition, they provide no link between the behavior of the basic mechanisms alluded to in the models and our daily-life visual experience.

This section describes a model of early spatial visual processing that overcomes the drawbacks just outlined. The present model characterizes early spatial visual processing in such a way that the perceived image corresponding to a scene fixated on a certain location can be reproduced. An experimentally measurable parameter in the model describes the fall of image quality with distance to the fixation location, and image processing techniques can then be used to transform the input image in order to show what its perceived version is like. In essence, this model of visual inhomogeneneous processing explains how "the whole configuration is not the effective stimulus for perception" (Peterson & Hochberg, 1983, p.183), provides us with a copy of what that effective stimulus looks like at a glance, and allows hypotheses to be raised about how its characteristics are related to the occurrence of perceptual phenomena.

Following an image-forming metaphor drawn by Schade (1956), visual scientists have since measured the performance of the visual system as an image-forming device by means of sinusoidal gratings. This methodology provides a unified basis for studying the perception of any image because any image can be expressed as a sum of sinusoidal gratings. Thus, knowledge of how the visual system processes each member of this set of basis stimuli can be used to describe the processing of complex stimuli as a combination of the processing of its components. This statement that visual processing of an image can be understood as the sum of the processing of its individual components (sinusoidal gratings) relies on the fact that the visual system is linear to a reasonable extent, although not completely so.

By measuring the threshold contrast needed for a human subject to detect foveally fixated sinusoidal gratings varying in spatial frequency and orientation, the visual system can be charaterized as a spatial-frequency filter. The empirical relationship of threshold contrast to spatial frequency leads to defining the *contrast sensitivity function*, which is the characteristic function of the visual system as a spatial-frequency filter. As it turns out, the foveal contrast sensitivity function covers a spatial frequency band ranging from 0 to about 60 cycles per degree of visual angle (c/deg), and on a logarithmic scale it has the shape of an inverted U, with a peak at about 4-6 c/deg. Therefore, the detectability of foveally fixated gratings depends on their spatial frequency: gratings of very high or very low spatial frequencies require more physical contrast than do gratings of intermediate spatial frequencies to be detected.

Sinusoidal gratings displayed on the face of a monitor for measuring the foveal contrast sensitivity function do not cover the entire visual field, and it was soon realized that the retinal area covered by the grating was an influential variable in threshold measurements. In fact, the foveal threshold contrast for detection decreases asymptotically with increasing retinal size of the grating (Hoekstra, van der Goot, van den Brink, & Bilsen, 1974). In addition, contrast sensivitity measurements off the fovea revealed that the peripheral contrast sensitivity function is displaced towards lower spatial frequencies the larger the distance to the fovea (Rovamo, Virsu, & Nasanen, 1978). These two facts suggest that the visual system be described as a *space-variant spatial-frequency filter.*

Garcia-Perez (1988) proposed a model of space-variant visual processing that accommodates these two empirical facts of visual performance at threshold contrast. The model assumes that the visual system contains local units or sensors that are selectively sensitive to a narrow range of spatial frequencies at their location. Sensors are grouped into sets or channels such that all members in a set share the same attributes (spatial-frequency selectivity) and differ only in location in the visual field; different sets of sensors differ in their spatial-frequency selectivity. Sensors in a channel are spatially distributed around the fovea within the limits of a circular area whose radius is inversely related to the spatial-frequency band the channel is sensitive to: high spatial-frequncey sensors do not extend very far away from the fovea, sensors tuned to the next lower spatial frequency extend a little farther away, and so on. This model was empirically tested and its critical parameters, namely, the area of the visual field that is subserved by each channel (i.e., its spatial extent), were estimated (see Garcia-Perez, 1988, experiment 1).

So far, only visual processing at threshold contrast has been considered. Since perception normally occurs at well above-threshold contrast levels, the visual processing to high-contrast images has to be considered separately. And, in fact, sinusoidal gratings which are useful for describing visual performance at threshold contrast can also be used for characterizing visual processing at suprathreshold contrast. The above model also handles this situation.

The distribution of sensors determines how much of the local information in the stimulus will be coded by the visual system at each location. At the fovea, where sensors exist for the analysis of all spatial frequencies, no information is lost. At peripheral locations, where high spatial-frequency sensors are not found, the visual system is blind to high spatial-frequency information. The loss of information increases with increasing distance to fovea and it results in an increasing blur of the perceived image with increasing distance to the fixation point. The spatial area where stimulus information in each band of spatial frequencies is visible in high-contrast images was also measured by Garcia-Perez (1988, experiment 2). Figure 1 shows a plot of the experimental results. The height of the intersection of a vertical line at a given abscissa describes the distance to the fovea up to which the corresponding spatial-frequency information is coded. An alternative, but equivalent, way to look at these data is by considering the horizontal segment connecting the curve with the vertical axis at a given ordinate. That segment refers to the range of spatial frequencies that are visible at the corresponding distance to the fixation location. Note that all spatial frequencies are visible within an area extending up to 2 degrees from the fovea, a result that bears on Nelson and Loftus (1980) finding that qualitatively different information is acquired from the region extending about 1.5 degrees from fixation than from any other farther region.

The space defined in the plot represent visual information that is present in the stimulus: information at all spatial frequencies exits everywhere in the stimulus. The experimental curve splits that space into two regions: that below the curve, which contains the information that will be coded by the visual system, and that above it, containing the information to which the visual system is blind. A copy of the perceived image of a stimulus given a fixation location can be obtained by simulating visual processing using image processing techniques to do away with the information that the visual system will not code.

A thorough description of that technique and its rationale can be found elsewhere (Garcia-Perez, 1991). Figure 2 illustrates its various steps. First, the stimulus image (Fig. 2a) is decomposed into several filtered images, each containing a separate band of spatial frequencies (Fig. 2b). Then, once the fixation location and the angle subtended by the image has been set, information outside the area of visibility in each filtered image is removed (Fig. 2c). Finally, the windowed images are combined to render the perceived image (Fig. 2d).

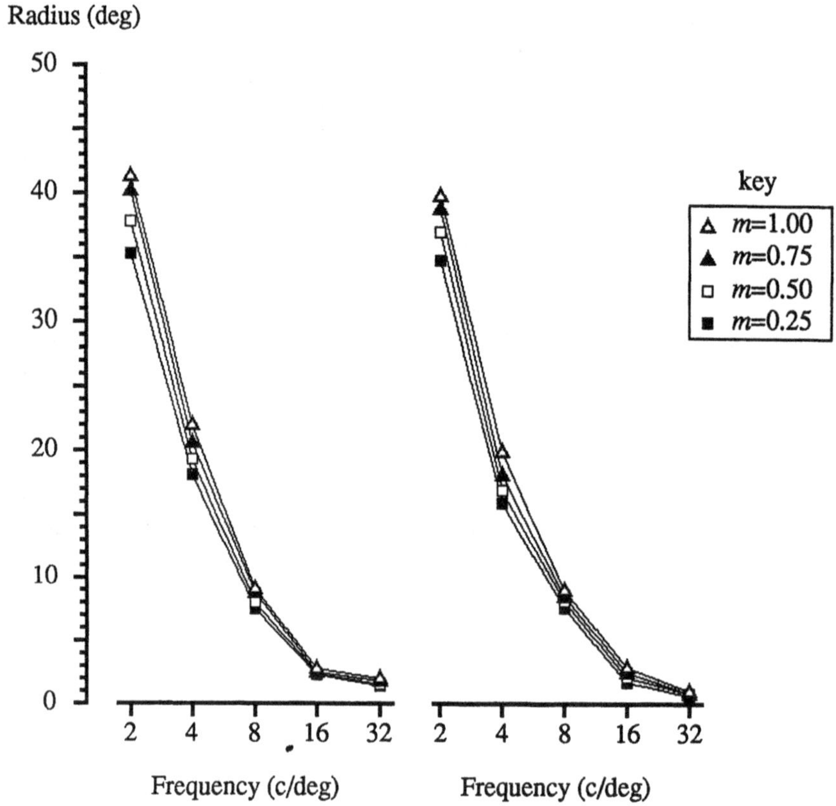

Fig. 1. Radius (in degrees of visual angle) of the area around the fovea where information at different spatial frequencies is coded. Data from two subjects are plotted in separate graphs. Each curve corresponds to a different value of well above-threshold contrast, *m*, as described in the inset. (Reprinted from García-Pérez, 1988; © 1988 by International Science Publishers, B.V.)

Fig. 2. Steps to obtaining the perceived image of a picture. The checkerboard in (a) is assumed to be fixated on the center, subtending 24x24 degrees. (b) Decomposition of the checkerboard into separate band-pass images, with bands centered, from top to bottom, at 32, 16, 8, 4, and 2 c/deg. The bottommost image is actually low-pass because it also contains the residual low spatial-frequency information. (c) Windowing of the band-pass images to eliminate spatial-frequency information falling outside the areas where it is coded by the visual system. (d) Perceived image computed by summing up the windowed band-pass images.

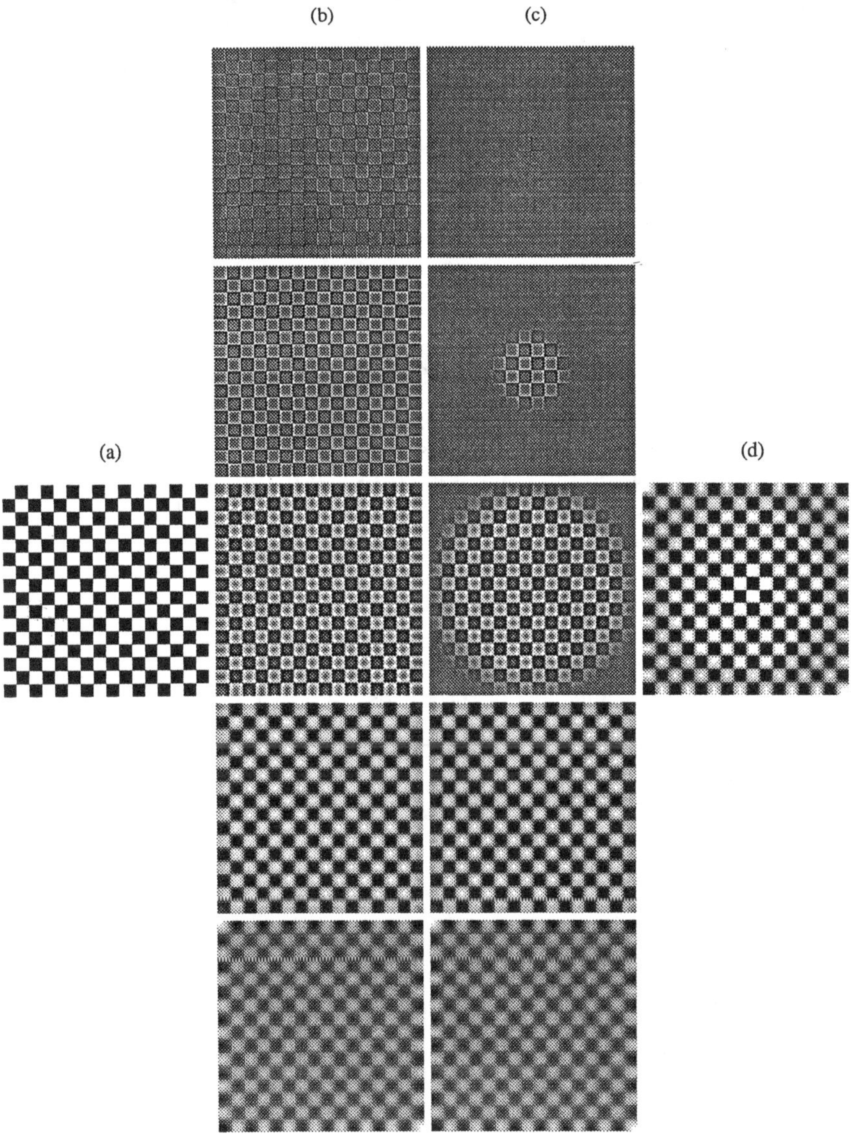

Fig. 2.

Given the qualitatively different information that is obtained from different locations in the visual field within a glance, the possibility exists of relating visual phenomena in general (and perceptual reversals in particular) to changes in the perceived image that are due to changes in the location of fixation.

5. The Role of Eye Movements in Solving Multistable Figures

Eye movement recordings during inspection of pictorial stimuli have provided some data relevant to the topic that is being dealt with in this chapter. It is now well established that human subjects tend to fixate certain parts of a pictorial stimulus, presumably in order to interpret the scene. It seems as though those parts of the scene carried enough information for the subject to start understanding the whole scene. Once a local area of the scene has been interpreted, the remaining parts of it are more readily interpreted according to, and with the help of, the constraints provided by the meaning assigned to the area interpreted first. Certain areas in a scene are more likely than others to boost this interpretational process. As yet we do not have a complete answer as to why this should be so.

When multistable figures are considered, different elements in them can start the process in a way that will lead to one interpretation or the other. If this process is iterative-which is likely to be the case when subjects know beforehand that the figure is subject to several interpretations- then a change in fixation location might give rise to a change in the starting point of the process and, hence, in its final outcome.

It will be assumed here that this interpretational process starts with an analysis of the available high spatial-frequency (fine detail) information. Evidence for the primary role of high spatial frequencies in picture perception comes from experiments showing that image edges are the critical features for recognition (Biederman & Ju, 1988). Therefore, fixations around a focal area for one of the percepts of the multistable figure will bring out fine-detail information that is relevant to that percept, and the interpretational process would be more likely to build it up. That eye movements, in conjuction with visual inhomogeneity, actually accomplish this information selection is shown in Figs. 3-7.

Figure 3a shows a version of Boring young/old woman (an ambiguous figure). The perceived image corresponding to a fixation midway between the young woman's eye and ear when the figure subtends 15x20 degrees of visual angle is

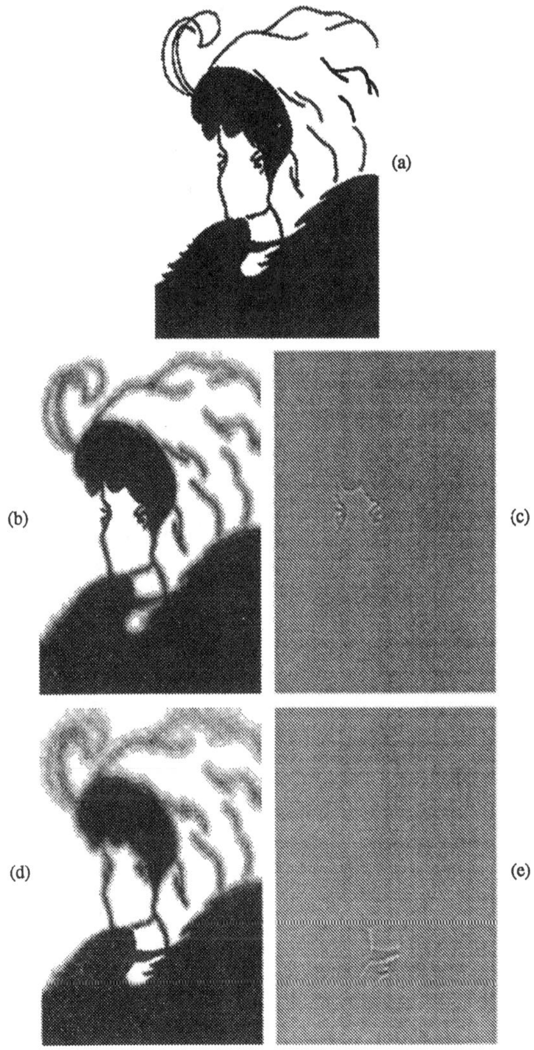

Fig. 3. (a) Boring's young/old woman. (b) Perceived image when the figure subtends 15x20 degrees of visual angle and fixation is midway between the young woman's eye and ear. (c) High spatial-frequency (above 16c/deg) information available under those viewing conditions. (d) Perceived image when fixation is on the old woman's mouth. (e) High spatial-frequency information available under such circumstances. (Reprinted from García-Pérez, 1989b; © 1989 by the Psychonomic Society, Inc.)

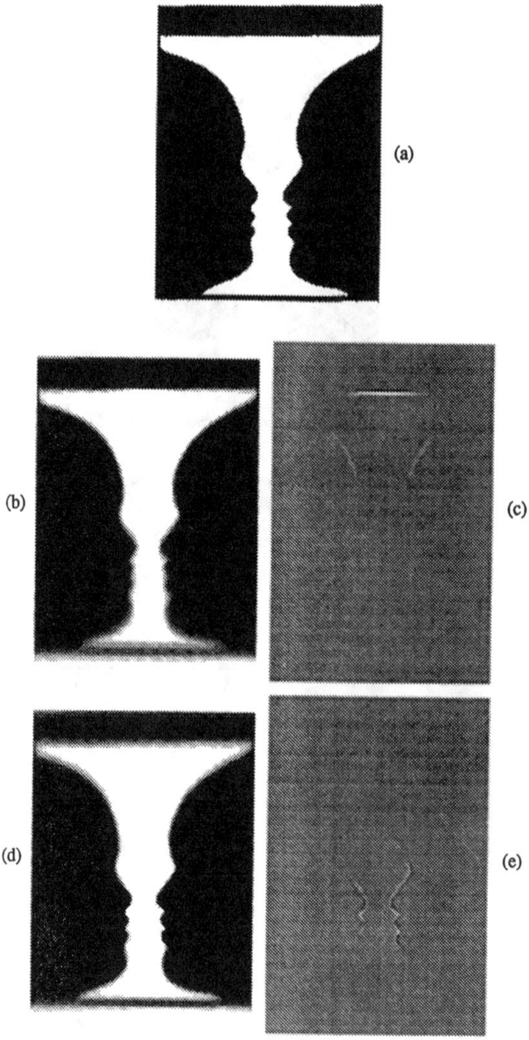

Fig. 4. (a) Rubin's vase/face. (b) Perceived image when the figure subtends 9x12 degrees of visual angle and fixation is on the center of the vase hollow. (c) High spatial-frequency (above 16 c/deg) information available under those viewing conditions. (d) Perceived image when fixation is on the nose of the right human profile. (e) High spatial-frequency information available under such circumstances. (Reprinted from García-Pérez, 1989b; © 1989 by The Psychonomic Society, Inc.)

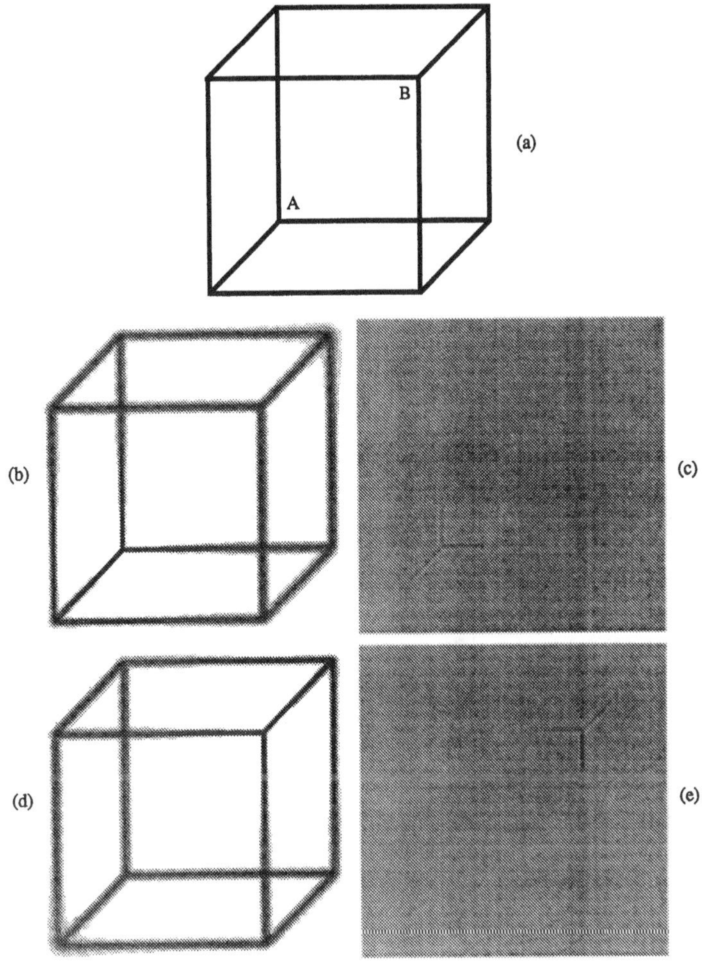

Fig. 5. (a) Necker's cube. (b) Perceived image when the figure subtends 10x10 degrees of visual angle and fixation is on the corner marked A in Fig. 5a. (c) High spatial-frequency (above 16c/deg) information available under those viewing conditions. (d) Perceived image when fixation is on the corner marked B in Fig. 5a. (e) High spatial-frequency information available under such circumstances.

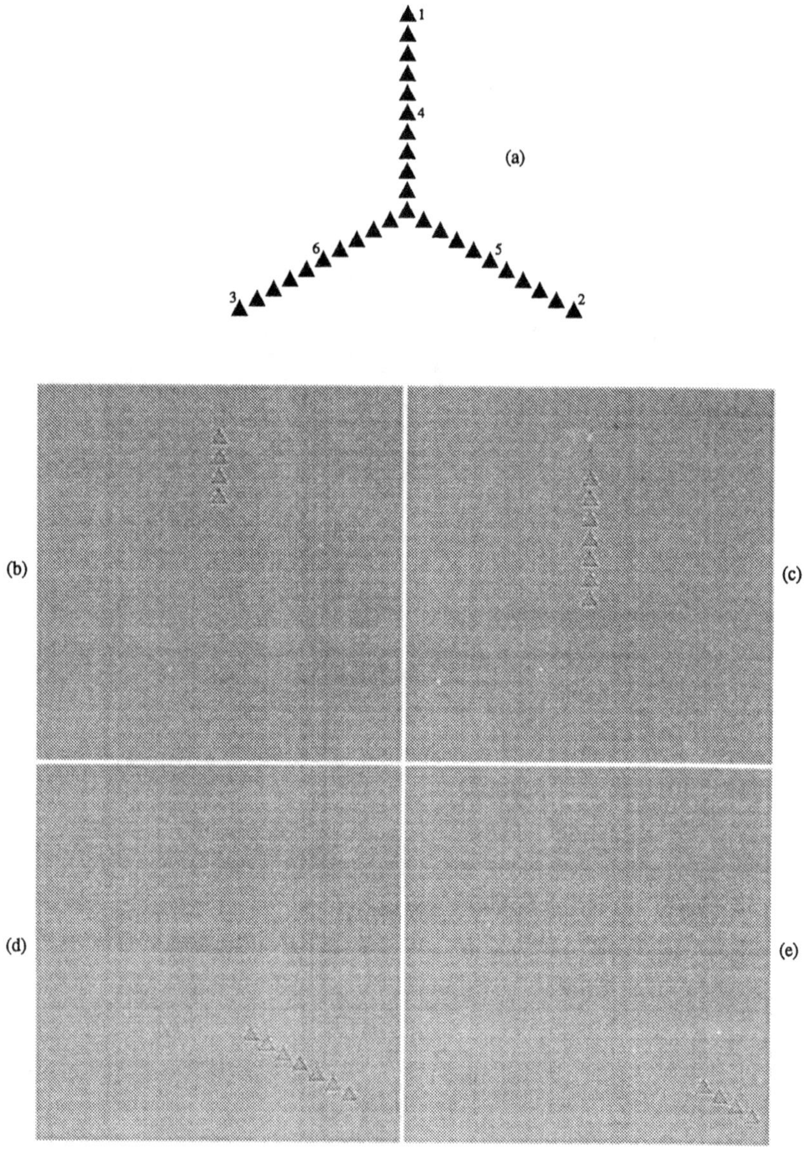

Fig. 6.

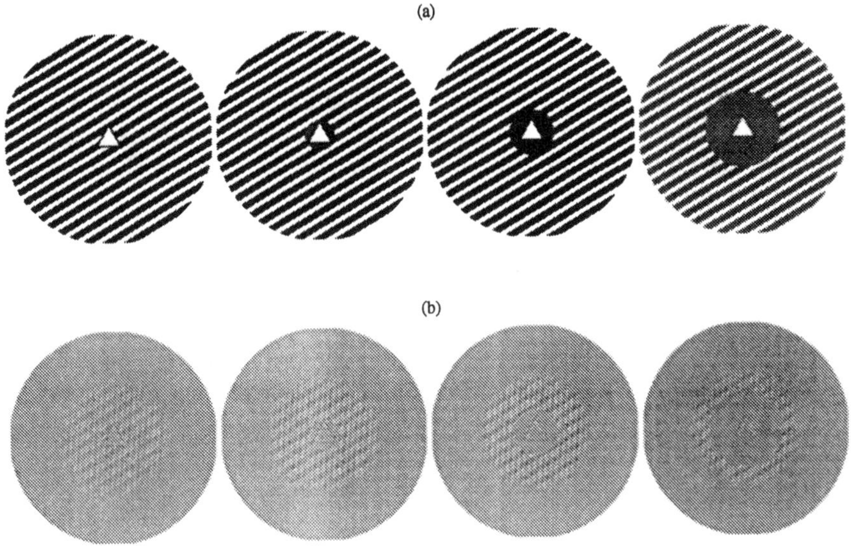

Fig. 7. (a) Orientation reversible triangles surrounded by a 2.7 c/deg square-wave grating patch whose inner edge is located at various distances to the center of the triangle. (b) High spatial-frequency information available when the triangles are fixated on their centers and each individual image subtends 7.2x7.2 degrees of visual angle.

Fig. 6. (a) Configuration of orientation reversible triangles. Some triangles are labelled for later reference. (b) High spatial-frequency information available when the configuration subtends 10x10 degrees of visual angle and triangle 1 is fixated. (c) High spatial-frequency information available when triangle 4 is fixated. (d) High spatial-frequency information available when triangle 5 is fixated. (e) High spatial-frequency information available when triangle 2 is fixated. (Reprinted from García-Pérez, 1989a; © 1988 by The Psychonomic Society, Inc.)

shown in Fig.3b, and the available high spatial-frequency information under such circumstances in shown in Fig.3c. This information is clearly confined to the focal area for the young woman interpretation, as reported by Gale and Findlay (1983). Figure 3b shows the perceived image when the old woman's mouth is fixated under the same angular subtense. The high spatial-frequency information that is then available (Fig.3e) is confined to the focal area for the old woman interpretation (see Gale & Findlay, 1983).

These same considerations apply to figure-ground reversible figures, as shown in Fig.4 for the Rubin vase/face (Fig.4a). Figure 4b and 4d respectively show the perceived images corresponding to fixations on the center of the vase hollow and the nose of the right human profile when the image subtends 9x12 degrees. High spatial-frequency information available in each case is shown in Figs.4c and 4e. In Fig.4b, the human profiles are too blurred to be found in the high spatial-frequency image and, therefore, their presence is likely to remain unnoticed at the time when the interpretational process starts. In Fig.4d, those profiles are so clearly apparent (and, at the same time, clues for the presence of a vase so absent) that the interpretation of two facing human profiles is immediate and almost unavoidable.

As for perspective-reversible figures, Fig.5 presents an analysis of fixations on Necker cube that is similar in all respects to that which was presented in Figs.3 and 4. Figures 5b and 5d respectively show the perceived images when points A and B in the cube in Fig. 5a, assumed to subtend 10x10 degrees, are fixated. Figures 5c and 5e show the high spatial-frequency information that is available in each case. If the vertex thus selected is to be interpreted as the corner of a solid object, then it is clear that a different three-dimensional solution for the cube has to be associated with each of these fixations.

With regard to orientation-reversible shapes, the limited availability of high spatial-frequency information sets out the basis for an explanation of the bias induced by the addition of configural and textural information, as described in Section 3.5 above. Figure 6a shows one of the stimulus used by Garcia-Perez (1989a). The subjects' task was to determine whether the direction of pointing of a specific triangle in the figure was up, down, left, or right. The triangle whose direction of pointing the subject had to report in each trial was signaled by presenting a fixation point immediately before presentation of the configuration. The target triangles were those labeled with numbers in Fig.6a. Results showed that response times to triangle 4 were the shortest, closely followed by response times to triangle 1; responses to triangles 2 and 3 were similar and much slower than responses to triangle 1; finally, responses to triangles 5 and 6, which were also similar, were the slowest. These results agree well with results obtained by Palmer and Bucher (1981) using other configurations, and can be cast in terms of the amount of facilitation or interference induced by the particular arrangement of the triangles in the configuration. Since the actual stimulus is the same regardless of where the subjects fixate, an explanation for these facilitating or interfering effects has to be found in the visual processing of the image rather than in the characteristics of the configuration itself. Figures 6b to 6e respectively show the high spatial-frequency information that is available when triangles 1, 4, 5, and 2 are fixated. Only part of the whole configuration is coded in the high spatial-frequency range, and the characteristics of the surviving information can account for the experimental data in terms of the parametric results obtained by Palmer and Bucher (1981): when triangles are aligned consistently with the required response (as in Figs.6b and 6c), the response time is short, decreasing further as the number of elements in configuration is increased; when triangles are aligned inconsistently with the required response (as in Figs. 6d and 6e), the response time is long, increasing with the number of elements in configuration.

The bias induced by textural context can be accounted for similarly. Palmer and Bucher (1982, experiment 5) reported that a background texture consisting of a 2.7 c/deg square-wave grating biases the perceived pointing of a triangle to an extent that depends on the distance between the fixated triangle and the inner edge of the textured background. Figure 7a shows versions of various of the stimuli used in that experiment, and Fig.7b shows the high spatial-frequency information available when each individual stimulus subtends 7.2x7.2 degrees of visual angle, with fixation on the center of the triangle. As can be easily seen, because the outer limit for the availability of high spatial-frequency information is determined by visual inhomogeneity, the amount of biasing high spatial-frequency information other than the triangle itself diminishes as the inner edge of the grating patch moves away from the triangle, and experimental response times in this condition of interference were found to vary directly with the amount of effective bias found in these images.

6. Summary and Conclusions

From the figures presented in the previous section, the differential blurring in the perceived image that is due to visual inhomogeneity can be interpreted as a sort of bias unavoidably introduced in the image by the visual system. Therefore, the perceived image of a multistable figure is always biased no matter how hard one tries to create a purely ambiguous figure. This bias, however, has a different sign depending on what location one actually fixates when looking at the figure. If the available high spatial-frequency information determines the interpretation of a figure (whether multistable or not) then, in practice, fixating on a given location of the figure (whether multistable or not) will be about the same as covering the figure with an opaque mask with a round hole that leaves visible only a small portion of it. Under those circumstances, if the visible part of a non-multistable figure makes some sense, then a picture of the whole figure will be more or less visualized. In case that the visible portion is unintelligible, then no meaningful perceptual response will occur, and a change in the location of the window will be sought. What characterizes multistable figures is that the visualized picture resulting from occlusion of certain parts of it differs depending on what parts are occluded and what others are left visible.

Under normal viewing conditions (i.e., without artificial occlusions), visual inhomogeneity performs this selection of information, and eye movements determine the particular area from which the information will be extracted at each point in time. Therefore, eye movements help interpret complex (non-multistable) scenes by allowing the visual system to pick out information from all locations to build up an internal representation (Baron, 1985; Julstrom & Baron, 1985). When eye movements occur while viewing multistable figures, then the possibility exists that a change in fixation location will give rise to a completely different percept instead of gathering more evidence in favor of the currently held percept.

It could be said against the points raised in this paper that it is visual attention and not eye fixations that determine the focused (metaphorically) part of the figure. It is known that while the locus of attention normally coincides with the locus of fixation, attention can be directed somewhere else (Posner, 1980; Tsal, 1983) and it is also well documented that visual search, which requires movements of attention, can occur in the absence of eye movements (Klein & Farrell, 1989). Whether private investigator or not, at least once in a lifetime everyone has fixated on some point while actually looking at what was happening somewhat off the fixation location out of the corner of one's eyes. This fact argues in favor of a lack of coincidence between the fixated spot and the one attended to . Despite this, what cannot be denied is that eye fixations supply optically focused information about the fixated parts of a scene. And, in the absence of a voluntary aim at concentrating on a location off fixation, that is, if attention is drawn to the fixated location, then the bias induced by visual inhomogeneity may be responsible for the perceptual reversals of multistable figures that follow a change in fixation location.

As Coren and Girgus (1978, p.565) said with regard to visual illusions, to talk of *the* cause for the alternation in perceptual multistability seems somewhat misguided. The argument developed in this paper accounts for *some* of the reversals: those that occur in conjunction with eye movements. And the way those reversals are accomplished is explained in terms of a kind of bias induced by visual inhomogeneity. But there is undoubtedly more to perceptual multistability than can be explained by resorting to a single mechanism or process.

References

Attneave, F. (1968). Triangles as ambiguous figures. *American Journal of Psychology, 81,* 447-453.

Attneave, F. (1971). Multistability in perception. *Scientific American, 225,* 62-71.

Baron, R.J. (1985). Visual memories and mental images. *International Journal of Man-Machine Studies, 23,* 275-311.

Biederman, I., & Ju, G. (1988). Surface- versus edge-based determinants of visual recognition. *Cognitive Psychology, 21,* 38-64.

Borsellino, A., Carlini, F., Riani, M., Tuccio, M.T., De Marco, A., Penengo, P., & Trabucco, A. (1982). Effects of visual angle on perspective reversal for ambiguous patterns. *Perception, 11,* 263-273.

Borsellino, A., De Marco, A., Allazetta, A., Rinesi, S., & Bartolini, B. (1972). Reversal time distribution in the perception of visual ambiguous stimuli. *Kybernetik, 10,* 139-144.

Braunstein, M.L., & Andersen, G.J. (1984). A counterexample in the rigidity assumption in the visual perception of structure from motion. *Perception, 13,* 213-217.

Bressan, P., & Vallortigara, G. (1986). Multiple 3-D interpretations in a classic stereo-kinetic effect. *Perception, 15,* 405-408.

Buffart, H., & Leeuwenberg, E. (1983). Structural information theory. In H.G. Geisler, H. Buffart, E. Leeuwenberg, & V. Sarris (Eds): *Modern Issues in Perception.* (pp.48-72.) Amsterdam: North-Holland.

Coombs, C.H. (1983). *Psychology and Mathematics.* Ann Arbor, MI: The University of Michigan Press.

Coren, S., & Girgus, J.S. (1978). Visual illusions. In R. Held, H.W. Leibowitz, & H.-L. Teuber (Eds): *Handbook of Sensory Physiology.* (Vol. VIII, pp.549-568.) Berlin: Springer-Verlag.

De Marco, A., Penengo, P., Trabucco, A., Borsellino, A., Carlini, F., Riani, M., & Tuccio, M.T. (1977). Stochastic models and fluctuations in reversal time of ambiguous figures. *Perception, 6,* 645-656.

Ellis, S.R., & Stark, L. (1978). Eye movements during the viewing of Necker cubes. *Perception, 7,* 575-581.

Epstein, W., & Rock, I. (1960). Perceptual set as an artifact of recency. *American Journal of Psychology,* 73, 214-228.

Fisher, G.H. (1967). Preparation of ambiguous stimulus materials. *Perception & Psychphysics, 2,* 421-422.

Gale, A.G., Findlay, J.M. (1983). Eye movement patterns in viewing ambiguous figures. In R. Groner, C. Menz, D.F. Fisher, & R.A. Monty (Eds): *Eye Movements and Psychological Functions: International Views.* (pp.145-168.) Hillsdale, NJ:Erlbaum.

Garcia-Perez, M.A. (1988). Space-variant visual processing: Spatially limited visual channels. *Spatial Vision, 3,* 129-142.

Garcia-Perez, M.A. (1989a). Visual inhomogeneity and reference frames. *Bulletin of the Psychonomic Society, 27,* 21-24.

Garcia-Perez, M.A. (1989b). Visual inhomogeneity and eye movements in multistable perception. *Perception & Psychophysics, 46,* 397-400.

Garcia-Perez, M.A. (1991). The perceived image: Efficient modelling of visual inhomogeneity. *Spatial Vision,* in press.

Girgus, J.J., Rock, I., & Egatz, R. (1977). The effect of knowledge of reversibility on the reversibility of ambiguous figures. *Perception & Psychophysics, 22,* 550-556.

Goolkasian, P. (1987). Ambiguous figures: Role of context and critical feature. *Journal of General Psychology, 114,* 217-228.

Harris, J.P. (1979). The Schroder Staircase: A new perspective. *Perception & Psychophysics, 26,* 312-318.

Hoekstra, J., van der Goot, D.P.J., van den Brink, G., & Bilsen, F.A. (1974). The influence of the number of cycles upon the visual contrast threshold for spatial sine wave patterns. *Vision Research, 14,* 365-368.

Howard, I.P. (1961). An investigation on a satiation process in reversible perspective of revolving skeletal shapes. *Quarterly Journal of Experimental Psychology, 13,* 19-33.

Julstrom, B.A., & Baron, R.J. (1985). A model of mental imagery. *International Journal of Man-Machine Studies, 23,* 313-334.

Kawabata, N. (1986). Attention and depth perception. *Perception, 15,* 563-572.

Kawabata, N., Yamagami, K., & Noaki, M. (1978). Visual fixation points and depth perception. *Vision Research, 18,* 853-854.

Kawamoto, A.H., & Anderson, J.A. (1985). A neural network model of multistable perception. *Acta Psychologica, 59,* 33-65.

Kelly, D.H., & Burbeck, C.A. (1984). Critical problems in spatial vision. *CRC Critical Reviews in Biomedical Engineering, 10,* 125-177.

Kienke, P.K., Sejnowski, T.J., Hinton, G.E., & Schumacher, L.E. (1986). Separating figure from ground with a parallel network. *Perception, 15,* 197-216.

Klein, R., & Farrell, M. (1989). Search performance without eye movements. *Perception & Psychophysics, 46,* 476-482.

Klymenko, V., & Weisstein, N. (1986). Spatial frequency differences can determine figure-ground organization. *Journal of Experimental Psychology Human Perception & Performance, 12,* 324-330.

Kohler, W., & Wallach, H. (1944). Figural aftereffects: An investigation of visual processes. *Proceedings of the Amercian Philosphical Society, 88,* 269-357.

Kolers, P.A., & Smythe, W.E. (1984). Symbol manipulation: Alternatives to the computational view of mind. *Journal of Verbal Learning and Verbal Behavior, 23,* 289-314.

Liebert, R.M., & Burk, B. (1985). Voluntary control of reversible figures. *Perceptual and Motor Skills, 61,* 1307-1310.

Lindauer, M.S., & Lindauer, J.G. (1970. Brightness differences and the perception of figure-ground. *Journal of Experimental Psychology, 84,* 291-295.

Long, G.M., Toppino, T.C., & Kostenbauder, J.F. (1983). As the cube turns: Evidence for two processes in the perception of a dynamic reversible figure. *Perception & Psychophysics, 34,* 29-38.

Luce, R.D. (1985). Mathematical modeling of perceptual, learning, and cognitive processes. In S. Koch, & D.e. Leary (Eds): *A Century of Psychology as a Science.* (pp.654-677.) New York: McGraw-Hill.

Masulli, F., & Riani, M. (1989). Ambiguity and structural information in the perception of reversible figures. *Perception & Psychophysics, 45,* 501-513.

Nelson, W.W., & Loftus, G.R. (1980). The functional visual field during picture viewing. *Journal of Experimental Psychology: Human Learning & Memory, 6,* 391-399.

Owen, L.A. (1985). Dichoptic priming effectis on ambiguous picture processing. *British Journal of Psychology, 76,* 437-477.

Palmer, S.E. (1980). What makes triangles point: Local and global effects in configurations of ambiguous triangles. *Cognitive Psychology, 13,* 285-305.

Palmer, S.E. (1985). The role of symmetry in shape perception. *Acta Psychologica, 59,* 67-90.

Palmer, S.E., & Bucher, N.M. (1981). Configural effects in perceived pointing of ambiguous triangles. *Journal of Experimental Psychology: Human Perception & Performance, 7,* 88-114.

Palmer, S.E., & Bucher, N.M. (1982). Textural effects in perceived pointing of ambiguous triangles. *Journal of Experimental Psychology: Human Perception & Performance, 8,* 693-708.

Palmer, S.E., Simone, E., & Kube, P. (1988). Reference frame effects on shape perception in two versus three dimensions. *Perception, 17,* 147-163.

Pelton, L.H., & Solley, C.M. (1968). Acceleration of reversals of a Necker cube. *American Journal of Psychology, 81,* 585-588.

Petersik, T. (1979). Optical defocusing reverses perceptual organization. *Perception, 8,* 225-228.

Peterson, M.A., & Hochberg, J. (1983). Opposed-set measurement procedure: A quantitative analysis of the role of local cues and intention in form perception. *Journal of Experimental Psychology: Human Perception & Performance, 9,* 183-193.

Pheiffer, C.H., Eure, S.B., & Hamilton, C.B. (1956). Reversible figures and eye-movements. *American Journal of Psychology, 69,* 452-455.

Phillipson, O.T., & Harris, J.P. (1984). Effects of chlorpromazine and promazine on the perception of some multi-stable visual figures. *Quarterly Journal of Experimental Psychology, 36A,* 291-308.

Posner, M.I. (1980). Orienting of attention. *Quarterly Journal of Experimental Psychology, 32,* 3-25.

Poston, T., & Stewart, I. (1978). Nonlinear modeling of multistable perception. *Behavioural Science, 23,* 318-334.

Reisberg, D. (1983). General mental resources and perceptual judgments. *Journal of Experimental Psychology: Human Perception and Performance, 9,* 966-979.

Reisberg, D., & O'Shaughnessy, M. (1984). Diverting subjects' concentration slows figural reversals. *Perception, 13,* 461-468.

Riani, M., Oliva, G.A., Selis, G., Ciurlo, G., & Rossi, P. (1984). Effect of luminance on perceptual alternation of ambiguous patterns. *Perceptual and Motor Skills, 58,* 267-274.

Riani, M., Tuccio, T., Borsellino, A., Radilova, J., & Radil, T. (1986). Perceptual ambiguity and stability of reversible figures. *Perceptual and Motor Skills, 63,* 191-205.

Rovamo, J., Virsu, V., & Nasanen, R. (1978). Cortical magnification factor predicts the photopic contrast sensitivity of peripheral vision. *Nature, 271,* 54-56.

Sakano, N. (1963). The role of eye movements in various forms of perception. *Psychologia, 6,* 215-227.

Schade, O.H. (1956). Optical and photoelectric analog of the eye. *Journal of the Optical Society of America, 46,* 721-739.

Shank, M.D., & Walker, J.T. (1989). Figure-ground organization in real and subjective contours: A new ambiguous figure, some novel measures of ambiguity, and apparent distance across regions of figure and ground. *Perception & Psychophysics, 46,* 127-138.

Sisson, E.D. (1935). Eye-movements and the Schroder stair-figure. *American Journal of Psychology, 47,* 309-311.

Stewart, I.N., & Peregoy, P.L. (1983). Catastrophe theory modeling in psychology. *Psychological Bulletin, 94,* 336-362.

Taylor, M.M., & Aldridge, K.D. (1974). Stochastic processes in reversing figure perception. *Perception & Psychophysics, 16,* 9-27.

Tsal, Y. (1983). Movements of attention across the visual field. *Journal of Experimental Psychology: Human Perception & Performance, 9,* 523-530.

Tsal, Y., & Kolbet, L. (1985). Disambiguating ambiguous figures by selective attention. *Quarterly Journal of Experimental Psychology, 37A,* 25-37.

Vickers, D. (1972). A cyclic decision model of perceptual alternation. *Perception, 1,* 1-48.

Wilton, R.N. (1985). The recency effect in the perception of ambiguous figures. *Perception, 14,* 53-61.

The Role of Eye Movements in Perceptual Processes
E. Chekaluk and K.R. Llewellyn (Editors)

THE ROLE OF EYE MOVEMENTS IN READING: SOME LIMITATIONS OF THE EYE-MIND ASSUMPTION

GEOFFREY UNDERWOOD and JOHN EVERATT

1. Introduction to the Eye-Mind Assumption

This discussion will focus upon our understanding of the reading process as it has been illuminated by observations of eye movements made during the comprehension of written language.

Our starting point will be with the assumption that the direction of our eyes indicates the contents of our mind. If the assumption is invalid, then studies of eye movements are of little interest to investigations of reading. The assumption is in general valid, but one special case of invalidity concerns the eye guidance mechanism. Current evidence suggests that text which is as yet unfixated can be used to guide the eyes to the location of the next fixation, suggesting that the material under fixation is not the only material which is processed.

Once the eye-mind assumption is accepted as a rough guide, rather than as a general assumption, we can turn to studies of fixation duration. The readers' eyes remain on a word (or part of word) until processing is completed, and so variation in fixation duration can tell us about the difficulty of processing. Variations in word and sentence difficulty produce variations in fixation durations, but there is a small amount of spill-over in the fixations following fixation upon a difficult word, again providing a special exception to the assumption of linkage.

Several eye movement experiments on sentence processing have been claimed to support a model of human parsing in which a single syntactically determinate analysis is constructed autonomously by recourse to parsing strategies in the face of syntactic ambiguity. In these experiments not only data concerning fixation duration but also the pattern of initial and subsequent (sometimes regressive) fixations is used as evidence in support of the model. More recent work confirms the link between fixation duration and patterning with on-line parsing, but suggests that other non-syntactic aspects of written language, such as punctuation and semantic information, can affect this process.

In addition to using eye movements to investigate variations in the difficulty of processing texts, we can also use them to investigate difficulties experienced by readers of varying ability. We shall consider variations in normal readers who read sentences in preparation for comprehension questions, with examples taken from one of the parsing experiments discussed earlier, and also provide an analysis of a trained speed-reader who also took part in this experiment. The remarkable performance of this reader also provides a difficulty for the general form of the eye-mind assumption.

2. The directions of the eyes and the contents of the mind

How can we know what a reader is thinking about? This discussion examines the relationship between attention to an event and performance, and questions the assumption which identifies the locus of attention with the direction of the eyes. It further questions the assumption which says that the information flowing through the reader's mind can be determined by observations of the readers eyes. The evidence used in the evaluation of these relationships concerns the visual processing which can be completed without foveal inspection. This evidence suggests that we can process information which is not being fixated and that we tend to fixate informative parts of a display, and leads to the subsidiary question of how our eyes are guided over a visual scene.

A strong statement of the relationships between attention and the direction of our eyes comes in the form of Just and Carpenter's (1980) eye-mind assumption:

> ". . . the eye remains fixated on a word as long as the word is being processed." (p. 330)
> and
> ". . . the eye-mind assumption posits that there is no appreciable lag between what is being fixated and what is being processed." (p. 331)

That is, the direction of the reader's eyes provides us with a measure of what is going through the reader's mind. The assumption could be interpreted to refer exclusively to the cognitive processes necessary during sentence comprehension because it was presented in the context of a discussion of eye guidance during reading. The eye-mind assumption considers that a fixation will continue until all of the cognitive processes activated by the fixated word have been completed. However, previous knowledge and previously fixated words are also expected to influence fixation durations (Carpenter and Just, 1983, p. 276). This qualification weakens the assumption, because what is being fixated is not necessarily what is being processed. A reader may spend an unusually long time gazing at a full stop at the end of a sentence while reflecting upon the meanings recently encountered or even upon some personal memories triggered by those meanings, and in this case the direction of gaze can tell us little about the contents of the reader's mind. Our aim is to demonstrate that the assumption can be challenged even further, with evidence that information ahead of fixation is also processed. If information from earlier fixations and from information not yet fixated can both influence the current fixation duration, then the measure of fixation duration will be an indication of the processing of past, present and future information. If this is the case, then the time taken to process a newly fixated word will only be partly indicated by the duration of the gaze upon a word, because past and future information will also be contributing to the time taken by processing. The eye-mind assumption would then be seen to be an unacceptably serial view of the order in which words are processed.

The eye-mind assumption has a close relationship with a second assumption which we can refer to as the eye-attention assumption, although the two should not be considered to be identical. Mind is taken here to refer to all current cognitive processes regardless of the direction of attention, and regardless of the current contents of consciousness. Attention is taken to be an active process of selection by which events may gain the scrutiny of consciousness. These are the definitions of "mind" and "attention" which will be used in the assessment of the two assumptions which relate eyes, mind and attention.

When we attend we may become aware, but this is only part of the ensemble of cognitive operations currently being performed by the mind. Accordingly, the eye-mind assumption may be valid independently of the eye-attention assumption. There are three possibilities here: both assumptions are valid, neither of them, or the eye-mind assumption may be valid while the eye-attention is false. The eye-attention assumption is a stronger version of the eye-mind assumption, in that if it is valid then the eye-mind assumption is necessarily valid. If the direction of gaze indicates current cognitive processes without indicating what it is that the viewer is attending, then the eye-mind assumption would be valid and the eye-attention assumption invalid.

If both assumptions are valid, then our eyes would indicate what is being attended, and no more than this would be processed. The only cognitive processes would be those processes which require attention. For this identity assumption to be true, there would have to be no evidence of unattended processing, and no processes of which we are unaware. There may be those who hold the view that those processes which gain attention, and only those processes, are those which are currently active. The identity assumption seems unlikely in view of the data reported and reviewed by Dixon (1981), Marcel (1983), and others, concerning the influence of preconscious stimuli, and the data reviewed by Allport (1979), Underwood (1982) and others, concerning the influence of unattended stimuli.

An example of the data which justify a preliminary doubt about the validity of the eye-mind assumption comes from a tachistoscope study in which readers named briefly presented line-drawings of familiar objects (Underwood, 1976). They knew the locations of the drawings, and because the displays were brief (60 msec) there was good reason to fixate them. The exposure duration was too short to allow an eye movement during the presentation. The dependent measure was the time taken to name the drawing, and this was found to vary according to the presence of a word printed to the right. When the word (e.g. "TREE") was related to the picture (a bird), then a slower response was obtained, in comparison with conditions in which an unrelated word, a non-word, or no word was presented. The word was described to the subjects as a distractor and to-be-ignored. They were not asked to report it, and there was no reason for them to look at it. A summary of the data from the experiment is presented in Figure 1.

Why should the word influence the time taken to name the picture? The experiment was initially presented in the context of the role of attention in word recognition - it provides evidence of recognition in the absence of attention - but it also provides evidence of reading without fixation. The word was presented to the right of fixation, and the selective effect of the relationship between the meaning of the picture and the meaning of the word implies that the meaning of the word has been processed. This is a special case of a challenge to the eye-mind assumption because the readers' eyes were not pointing directly at a word which was processed. Fixation upon a word is not necessary for recognition of the word or for processing to the point where it could interfere with the process of recognising the picture, selecting the appropriate name for the picture, and articulating that name. This experiment established the case for parafoveal semantic processing, and resulted in speculations about the generality of parafoveal processing during normal reading (e.g. Underwood, 1981). If readers are influenced by words in the parafovea of their vision in laboratory studies, then perhaps they process parafoveally available information when reading text. This possibility raised a number of questions: is it possible to demonstrate parafoveal

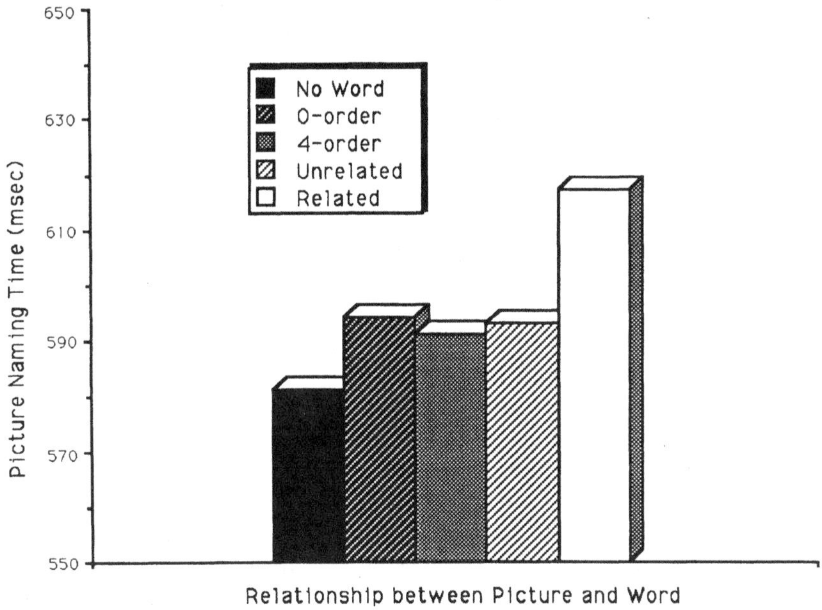

Fig. 1. Parafoveal words are recognised
Data showing an effect of an unattended, parafoveally presented word upon a picture naming task (Underwood, 1976). The nearest letter of the word appeared 1.5° from the centre of a simple line-drawing of a familiar object, and the task was to name the object as quickly as possible. The picture and word appeared simultaneously, and were shown for 60 msec. Subjects were instructed to ignore the word, but when picture and word were associated in meaning then there was a slower naming response. The five relationships between picture and word were as follows.

Related: The word was an associate of the picture (eg, picture of a *chair*; word – "TABLE");

Unrelated: The word was not normatively associated to the picture (eg, *boat*; "JUICE");

4-order: The picture was accompanied by a string of letters which had a fourth-order approximation to English (eg, *car*; "ERINC");

0-order: The picture was accompanied by a string of letters which had a zero-order approximation to English (eg, *pan*; "EJUDV");

No word: The picture appeared on an otherwise blank field.

processing in tasks where readers can move their eyes over a text rather than having single words displayed tachistoscopically, and what would be the purpose of this processing? To enrich the understanding of the text, perhaps, but if we can understand without fixating, this begs the question of why we need to fixate at all. The more likely suggestion was that the processing of parafoveal words was used by the eye guidance system, either positively, to direct the eyes to certain locations, or negatively, to guide the eyes away from some locations. These questions have provided the basis for a number of our studies, and the answers to some of them will be discussed in this chapter.

For our present purposes the early tachistoscope studies provide our first doubt over the validity of the eye-mind and eye-attention assumptions. They demonstrate that mental processing can, under admittedly special circumstances, proceed without attention and without fixation. If neither the eye-mind nor the eye-attention assumption are valid, then our eyes would never given an indication of current processes. There is abundant evidence to show that fixation patterns are sensitive to the difficulty of processing, and so this possibility may be rejected. The task is to describe the conditions under which the assumptions do hold.

Although Just and Carpenter are thinking specifically of the pattern of eye-movements which is observed during reading, the eye-mind assumption is a clear statement of a possible relationship between cognitive processing and the direction of gaze. While we are processing a stimulus, our eyes remain upon it. Some of this processing will involve recognition, and some may involve integration with previously seen stimuli, but the assumption suggests that there is no lag between what is being inspected and what is being processed. Similarly, there is no lag between what is being processed and what is being inspected. The mind is assumed to be neither ahead of, nor behind the eye. Eye-movement recordings can be informative only if the direction of the readers' eyes provides an indication of the cognitive processing of the text. The information provided by these recordings will be considered later in this review, as part of the discussion of how the skilled reader knows where to look next.

3. The case for the assumption

The amount of visual attention given to a word - the total amount of time the reader spends looking at it - varies according to the difficulty of recognition. Difficulty itself can be varied in a number of ways, but whenever we know a word to take longer to recognise, it is a safe generalisation to say that the word will receive more visual attention. This justifies the general case of the eye-mind assumption, and Just and Carpenter (1980) supported its use with data showing that during the reading of paragraphs taken from scientific texts the length of an inspection is directly related to the difficulty of processing. It is very well established that variables such as word length, frequency and novelty are amongst the strongest predictors of the duration of inspection. For example, one of their subjects looked at the relatively frequent word "question" for a period of 300 msec and the equally long but much less frequent "transfer" for 633 msec. Other investigators have reported similar effects. Long, unusual words gained very long inspections, with "thermoluminescence" being inspected for a total of 2431 msec whereas the average for all fixations across the course of the experiment was a gaze duration of 239 msec. If cognition is inevitably locked onto the direction of gaze, then we have a useful overt measure of our covert cognition processes. If the stronger eye-attention assumption is correct, then we also have a simple method of determining the direction of visual attention.

Does the mind process only that which is fixated? The eye-mind assumption claims that the eye remains fixated upon a word until processing has been completed, that there is no dissociation between what is being fixated and what is being processed, and a corollary is that cognition is determined by fixation. Whereas Just and Carpenter (1980) supported the assumption with data which shows that the time spent looking at a word is closely related to the difficulty of processing that word, Carpenter and Just (1983) presented further tests of the assumption by looking at the influences of the word preceding fixation, and the influences of the word ahead of fixation. If cognition is locked to fixation, then there should be no influence of material prior to or ahead of fixation.

There should be an influence of the word preceding the fixated word if processing lagged slightly behind the eye. If processing did lag behind the eyes, then the duration of the fixation upon one word would be a function of the difficulty of processing the word immediately to the left. Carpenter and Just found that the "gaze duration" received by a word was not affected by the length or frequency of the preceding word and, furthermore, this result holds whether the preceding word had been fixated or skipped.[1] The length

1 The use of gaze duration vs. fixation duration as the most representative measure of visual attention is not the subject of the present discussion. Suffice it to say that it forms the basis of another debate. Gaze provides an overall measure of attention while a single fixation duration gives a finer-grained estimate of moment-to-moment processing: if one wants to know whether to carry a raincoat out when taking a walk the monthly rainfall figures may not provide the best evidence. As a cumulative measure of all fixations "gaze" may simply be too insensitive to pick up all the effects of text processing upon eye guidance.

and frequency of the preceding word do not affect gaze duration, but we do not know about the influence of higher level linguistic processes such as syntactic assignment or propositional integration.

If attention moves ahead of the eyes then the word ahead of fixation might be expected to influence gaze duration, but Carpenter and Just found no effects of the length and frequency of the next word on the page. Regardless of whether the next word was eventually fixated, there was no effect of these variables, and this analysis means that we do not usually encode words to the right of the fixated word. We can conclude that the eye-mind assumption gives a good general description of the relationship between what is the direction of gaze and what is being processed. In the Carpenter and Just analysis there was, in fact, one very interesting effect of a content word ahead of fixation upon the current fixation of a function word, and, as we shall see, this effect favours a view which again challenges the generality of the eye-mind assumption.

The assumption is supported by good evidence: the difficulty of encoding of a word is the best predictor of the duration of gaze, and the encoding of the words immediately preceding and following the word have no effect upon this duration. This conclusion is not supported by evidence of non-fixated events being processed, such as in the early tachistoscope studies, and it is not supported by other data presented by Carpenter and Just. Although fixation duration appears to be unaffected by non-fixated words, there is good evidence that fixation location is determined in part by the distribution of information in the text. They reported that readers skip 18% of the content words and 62% of the function words. Hogaboan (1983) also reported that about 40% of words are not fixated when adults read coherent texts.

Why should readers fail to fixate words, and what can they know of these words? When skipping shows a sensitivity to word-type, then we have a further suggestion of processing without fixation. How could they know to not fixate a particular word if they did not know something about it? An important experiment reported by Fisher and Shebilske (1983) confirmed that a high proportion of words are not fixated, but that they were processed even when they are skipped.

The subjects were tested as yoked pairs, with the first member of each pair having their eye-movements monitored while reading. The words not fixated by this subject were not presented in the version of the text presented to this subject's partner. If the first subject fixated every other word, then these would be the words shown on the screen to the second subject. If the first subject fixated only one word in a sentence (as could happen with a speed reader, for instance), then the screen would be blank but for this one word when the second subject was tested. After presentation of the full text (for the first subject in each pair) or the impoverished text (for the second subject) there was a recall test. Of specific interest was the fixation and recall patterns of certain target words. Most of the time these targets were fixated, but on 12% of occasions they were skipped, with no fixation upon the word or within half a degree of it. In the recall test 59% of these non-fixated target words were available. The subjects' partners could only predict (or "recall") 26% of the non-presented words, indicating that the eye-movements subjects were relying upon more than textual context. The subjects who had to rely upon context to guess the target words were only half as accurate as the subjects who had the words available but failed to inspect them directly. When the eye-movements subjects did fixate the target word, their recall was still at 59% correct, suggesting perhaps that fixation directly upon a word is ineffective in determining retention in memory: fixating the target words did not provide an improvement in recall. Fisher and Shebilske concluded that their results directly opposed the predictions of the eye-mind assumption, and supported the notion that skilled readers make use of parafoveal vision. A similar conclusion about the parafoveal processing of skipped words was drawn by Slowiaczek (1983), and the importance of fixation probability is further discussed in the section on guidance models.

How do we know to not fixate a word without knowing that we have extracted a criterion amount of information through parafoveal processing? The reports of non-fixations during reading, and of the effects of these non-fixated words upon subsequent performance, suggest that the eye-mind assumption is at best a general description of reading behaviour, but that there are a number of special cases of processing without fixation. It may be the case that parafoveal processing is conducted without the direction of the reader's attention, and that the eye-attention assumption can be salvaged. In the following section the discussion turns to the relationship between the eye-mind assumption and the eye-attention assumption.

4. Attention and the eye-mind assumption

Before considering the eye-mind assumption in the context in which it is presented, it is appropriate to mention its boundary conditions. The direction of gaze tells us little about the contents of the mind, for instance, which the primary sensory input is other than visual. Only rarely will the direction of a conversationalist's eyes provide information about the content of his next utterance, and the direction of the listener's eyes will rarely tell us about his interpretation of the message being received. Even in the more relevant case of reading, my eyes can give a very false impression of my comprehension processes in the special instance of distracted reading. If I am distracted by some non-textual thought then the appearance of my eyes moving over the page may deceive an observer into believing that textual comprehension is occurring. In this instance I may be contemplating some future plan or some past memory, but my mind is not concerned with the text. My attention is not directed toward the input provided by my moving eyes, but toward my reminiscences or my plans. As my eyes arrive at the bottom of the page I may realise that I have not been thinking about the sentences, and that a second reading is necessary. (This is really the first reading, in the proper sense of reading as interpretation or comprehension). To my knowledge there have been no investigations of this phenomenon, and so it is not possible to say whether the eye-movements of the distracted reader differ from those of the concentrating reader, but either way the direction of gaze will not inform the observer of the content of the reader's mind.

Another special challenge to the eye-mind assumption comes from the case of the line of gaze providing ambiguous evidence of the direction of attention. When two objects are in close retinal alignment but at different distances, then attention can be switched between them without any noticeable change in the direction of gaze. For example, attention shifts without movement can be obtained with a word written on an otherwise clear pane of glass, with an object placed behind the window. Without any movement of the eyes attention can be alternated between the word and the object behind. This observation has been investigated formally by Kolers (1969), who designed a helmet-mounted mirror system for the purpose. A half-silvered mirror in front of the subject's eyes allowed for viewing of the scene forward, and reflected onto this mirror was the scene behind. The effect of the arrangement was that the subject could view the scene forward or behind without moving his eyes or his head - the scene behind was superimposed onto a natural view of what would be visible normally. Kolers placed objects and people in front and behind the subject, at similar distances, and asked for attention to be directed to one location or the other.

This switching of attention was described as being trouble-free and easy to accomplish.

The ease of moving attention without any movement of the eyes, which is demonstrated with Kolers' apparatus, provides a specific restriction upon our acceptance of the eye-mind assumption. Here is an instance, albeit an unusual one, of the direction of attention not being indicated by the direction of gaze. The line of gaze can sometimes provide ambiguous evidence. Of more interest, perhaps, are those instances where the direction of gaze provides misleading evidence of the direction of attention.

Whereas a form of the eye-mind assumption may be acceptable, the eye-attention assumption seems a little too strong for our intuitions. Helmholtz (1866) suggested that attention to a visual field can be concentrated away from the point of fixation, and Williams James (1890) also entertained the possibility of attending to objects in the periphery of vision whilst keeping the eyes still. This conclusion has been supported by recent empirical evidence, and the discussion will now turn to demonstrations of the ability to look at one feature in visual space while attending to another, with the aim of identifying further qualifications of the eye-mind assumption.

5. Attention and the direction of gaze

The most discriminating part of the retina is a small area - the fovea - and it is this densely packed area of cone receptors which is usually brought into alignment when patterns are inspected. As the distance of an image from the fovea increases, so our visual acuity for that image decreases, but the area of the fovea is not well defined, because there is a gradient of acuity. A safe working assumption is that the fovea has a diameter of about two degrees, and this gives maximum sensitivity to eight or ten printed letters when a book is read at a comfortable distance of, say 50 cm. It is this restriction upon the area of sensitivity which results in our saccadic eye movements during reading. When the fovea is aligned with a pattern, then we are considered to be fixating that pattern, and, for Just and Carpenter (1980) at least, it is this pattern which is being processed.

The skilled classroom teacher is reputed to have the ability of looking at one child while attending to the activities of another. This talent of "having eyes in the back of one's head" has clear survival value in a hostile classroom environment, but it is not clear whether attention can be sustained upon a non-fixated source, or more simply attracted by changes in the peripheral visual field. There have been a number of laboratory investigations of the ability to dissociate the point of fixation and the point of attention, and these will be considered here in the context of what they say about the eye-mind assumption.

We can inspect patterns without making eye-movements, and this has been demonstrated in experiments which project stabilised retinal images upon the retina. Although the eyes can move in these experiments, the image on the retina does not, and so the fovea cannot be brought to fixate a chosen part of the display. Secondly, the extraction of information from non-fixated parts of the visual field during and after tachistoscopic presentations also provides an instance of extra-foveal processing. Such presentations exclude eye-movements during inspection by using brief displays. As saccades do not start until about 180 msec after the instruction to move, any display which lasts for an interval less than this period of 'saccadic delay' will necessarily be shown to a stationary eye. Although these two types of experiments can be described as providing "special circumstances", they do serve to force a cautious interpretation of the eye-mind assumption. Even when the observer is free to move his eyes the non-fixated field can be processed: although we have greatest acuity for patterns projected to the fovea, we are not without sight in the periphery of vision.

6. Moving attention without eye movements

One of the most informative demonstrations of the dissociation between foveal fixation and attention is that reported by Grindley and Townsend (1968). Their experiments observed the detectability of a pattern which was presented to a non-fixated part of the visual field. The eye-movements of the subjects were monitored during the experiment, and only when the subject fixated the centre of the display was the trial included in the analysis. The pattern was an outline drawing of the letter T in one of four orientations and in one of four locations. It was displayed about 13 degrees away from the point of fixation, and the task was to report the location and orientation of the pattern. There were two further manipulations: one concerned an attention-directing cue prior to presentation, and one concerned the presence of distracting non-targets. The attentional cue was information about the location of the target, and this was always valid information presented immediately prior to the display. After its presentation the subject indicated that they were then attending the target location, without any change in eye fixation, and then the target was displayed.

The interesting result from Grindley and Townsend's experiment is that the attentional pre-cue had no effect upon the success of reporting the orientation of the target unless non-targets were present. If the target appeared against an otherwise blank field, then there was no advantage in directing attention to its location in advance. However, if the other three possible target locations were occupied by distractors, then it was useful to know where the target would be. Although it was presented 13 degrees away from the point of fixation, the pre-cue was able to provide a gain in performance. Almost twice as many targets were correctly identified with the pre-cue than when advance information was not given. Selection was necessary in the distraction condition, and it was here that the selective direction of attention gained benefits. This experiment not only provides a question mark over the eye-attention and eye-mind assumptions, in that it demonstrates that attention and pattern processing can proceed away from foveal fixation, but it also provides further information about the nature of the attention process. The preparation which accompanies the direction of attention to one part of the visual field is, in this experiment at least, preparation for filtering (Broadbent, 1971, 1982). Attention reduces the otherwise harmful effects of unwanted stimuli and the target is filtered through to the pattern recognition processes. When no distractors are present it is not necessary to select the target from among other patterns, and so no advantage is gained by pre-directed attention.

Grindley and Townsend (1968) demonstrated that attention can be allocated to peripheral vision when observers receive information about impending events of interest. As a way of determining more exactly the benefits which can be gained by directing attention in advance of a signal, an experiment by Posner, Nissen and Ogden (1978, Experiment 2) sometimes presented cues which were invalid and which therefore directed attention away from the anticipated target. On the other half of the trials a directional cue was presented in advance of the target, and this was an arrow pointing to the left or to the right. The arrow gave information about the timing and about the probable position of the target. This cue was intended to direct the observer's attention to one of the target locations and correctly predicted the location on 80% of trials. On the other trials it sent attention in the opposite direction.

Eye movements were monitored in the Posner et al. experiment, and so it is possible to ask whether observers can move their attention while maintaining their line of gaze. The benefits of moving attention to the correct location of the target can be determined by comparison with the neutral condition, in which no locational information was presented. The costs of moving to the wrong location can also be determined by comparison with the neutral condition. In terms of the saving in reaction time in comparison with the neutral condition, the size of the attentional benefit was approximately the same as the size of attention cost. By moving attention to the correct location of the target before the target appeared, a faster response could be made, but if attention moved in the wrong direction, then a slower response was made. Again these shifts in the direction of attention, which result in performance differences, are shifts of the mind without shifts of the eyes.

7. Moving attention and eyes independently

Attention can be dissociated from foveal inspection, but the experiments considered so far each demonstrate dissociation for the brief interval that it takes to present a single stimulus, and with stationary eyes. A series of experiments reported by Roger Remington (1980) carried the investigation of dissociation further, to observe the independent time course of eye and attention movements. The data suggested that whereas attention may move independently of the eyes, when the eyes do move they take attention with them. These conclusions have been confirmed by similar experiments reported by Shepherd, Findlay and Hockey (1986), who also found that it is not possible to make an eye-movement without making an attention-movement.

In Remington's first experiment the subjects fixated a cross in the centre of the screen, in anticipation of a target cross appearing to the left or the right. There were two possible target locations to the left, and two to the right, with the furthest location being 10 degrees from the centre of fixation. The fixation cross and the four possible target locations were indicated by empty squares which were displayed throughout the trial. Upon presentation of the target cross the subjects were to refixate upon it. Between 0 msec and 500 msec after onset of the target a second event sometimes occurred, and this was a brief increase (for 3 msec) in the brightness of one of the squares. Part of the task was to say whether or not there had been a brightness increment during the trial, and accuracy was recorded as a function of the relationship between the timing of the increment and the timing of the saccade. Remington found that the best detection of the brightness increment was for squares at the target location, presumably showing that if attention has been attracted to a location by the appearance of a target cross, then other events at this location will have a perceptual advantage. The interesting result was that this relationship held even for those increments which occurred before a saccade. On these trials the eyes would be fixating the central cross, and yet increments at the target location were detected better than increments of the square at the location of fixation. Here is a counter-intuitive result showing that we can sometimes see better when we are not looking at an event than when we are.

The dissociation found in Experiment 1 was repeated with a simpler display in Experiment 2, and then reversed in Experiment 3. The first two experiments reported that targets appearing to fixation were reported less well than those appearing at a peripheral location when attention was at that location, and the movement of attention and the eyes were both thought to be triggered by the appearance of a peripheral cue. In Experiment 3 the observers were asked to redirect their attention according to the direction suggested by a centrally fixated arrow which pointed to the left or to the right. There were single squares on each side of the central square (as in Experiment 2), and these were 9 degrees away from fixation. Remington considered that if shifts of attention were triggered peripherally, then they might not be made when the peripheral cue was absent, even though a saccade was being prepared and executed. Subjects refixated when the arrow appeared. Refixation took about 100 msec more with this central cue than with the peripheral cue, a result which confirms the attention-attracting potency of peripheral events. In addition, there was a tendency in this experiment for an easier detection of brightness increments at the central square, up to the time of the saccade. This contrasts with Experiments 1 and 2, in which a peripheral cue acted to improve the detectability of brightness increments in the periphery at this time. Shifts of attention are not made prior to a saccade if the movement is initiated by a centrally fixated cue.

The final experiment in the series asked subjects to maintain their attention in the centre of the screen while their eyes made a saccade to one side. With essentially the same displays as in Experiment 3, Remington instructed his subjects to move their eyes to the square in which a cross was to appear. On 80% of the trials, however, the brightness increment would occur in the central square which was, of course, the location of the initial fixation. Subjects were informed of this probability, and testing was sufficiently extensive to encourage them to attempt to attend to the central square regardless of the position of their eyes. A positive result in this experiment would have been evidence of a double dissociation between attention and fixation. A peripheral movement cue was used. Although it was in their interest to keep their attention on the central square, the subjects were only able to do this up to the onset of the saccade. After the saccade had been made, brightness increments at the new fixated location were more detectable than those at the central location. The sequence appears to be that attention is first focused on the central square, and then shifted to the target square at about the time of the saccade. By comparing Experiments 2 and 4 we can see that changing the distribution targets has the effect of holding attention on the more probable target location for a short time, but that as soon as the eyes move then attention moves also.

Remington's experiments showed a dissociation between eye and attention, but demonstrated that the time-scale of dissociation is small. Attention may momentarily jump ahead of an eye-movement when the peripheral display changes, and if the saccade is prompted by a symbolic instruction then the eyes may arrive first. The lag is small in each case: it is in the order of a tenth of a second at its maximum, and is possible to argue that the existence of such a lag contradicts the letter rather than the spirit of the eye-mind assumption. It certainly contradicts the eye-attention assumption, which requires a close locking of eye and attention.

The lag between eye and attention was in the order of one tenth of a second, and the question may be raised about the significance of such a small dissociation. It is worth pointing out, however, that in the experimental study of reading a tenth of a second is a long interval. In an associative priming experiment, for example, we have reported a longer fixation upon an unprimed word during sentence comprehension (Kerr and Underwood, 1985). The advantage of primed over unprimed words was 40msec of fixation duration, a value considerably less than the eye-attention lag found by Remington. These data are consistent with the hypothesis that attention can move ahead of the eyes, and facilitate pre-fixational processing of the text. If attention can be allocated to a word prior to fixation, then the question becomes one of how we know where to attend next - is our attention, and our eyes, guided to textual features or by them?

We can demonstrate that attention can be momentarily dissociated from the direction of gaze, but if we are to challenge the use of the eye-mind assumption in the study of reading, then it is necessary to demonstrate that, during reading, some features of text can be processed when they are not gaining foveal inspection. One problem for the assumption would arise if it were possible to demonstrate that under some circumstances the eye is drawn to a feature of text, for this would be a demonstration of pre-fixational processing. The following section of the discussion turns to the cognitive mechanisms which control eye guidance during reading.

8. Guidance models: How do readers know where to look next?

Three main models of eye-movement guidance recur in the literature in different guises. They can be ordered quite simply in terms of the amount of control exerted by the cognitive system of the reader, from models which suppose that our eyes move autonomously over a page, to those which suppose that our eyes selectively fixate features of the text according to the contribution those features can make towards the current comprehension calculation.

The Minimal Control Hypothesis (also known as the 'No Control' or 'Constant Pattern' or 'Oculomotor Control' model of eye guidance) proposes that the text has no direct influence upon the location of fixations. By this model, recommended by Tinker (1958), the eyes are preprogrammed to move a constant distance along the line. Variations in saccade length occur as a function of the reading skill of the individual, or as a function of the anticipated difficulty of the text to be read. A plausible variation of this model suggests that the current difficulty of the text can have an indirect and general effect upon eye-guidance by decreasing saccade lengths (and increasing fixation durations) when the cognitive load of the comprehension calculation increases. This variation is the Process-Monitoring or Gain Control Hypothesis, and has been favoured by Bouma and de Voogd (1974) and Shebilske (1975).

The Visual Control Hypothesis (also known as the 'Stimulus Control' or 'Peripheral Search Guidance' or 'Physical Aspects' model of guidance) proposes that the reader's eyes can be guided by the non-linguistic features of the text ahead of fixation (Hochberg, 1970). Purely visual features, such as the spaces between words and the presence of capital letters, might be detected in the parafovea or periphery, and used to guide the eyes towards informative locations on the page.

The Linguistic Control Hypothesis (also known as the 'Internal Control' or 'Cognitive Search Guidance' or 'Strong Control' model of guidance) is the most controversial of the three models. The reader's eyes are said to move to the next fixation according to the state of the comprehension calculation and according to syntactic or semantic variations in the text ahead of the eyes. Alternative forms of this model have been proposed by Hochberg (1970), Fisher (1975), Kennedy (1978) and Underwood (1981, 1985). Linguistic control may be exerted by the decoding of the text requiring the verification of predictions about the propositional structure. These predictions may require the inspection of certain words or certain parts of polymorphemic words. The strongest version of this hypothesis suggests that the eyes are attracted to linguistically informative parts of the text following parafoveal preprocessing, and this hypothesis continues to find empirical support (e.g. Fisher and Shebilske, 1985; Underwood, Clews and Everatt, 1990).

One point to note regarding these control hypotheses is that they are not mutually exclusive. A reader may rely principally upon visual information about word shape and word length, but may come under the influence of linguistic information if comprehension becomes locally difficult. Similarly, fixation locations may be determined according to the general difficulty of the text, in line with a process monitoring view, but with fixations occasionally being attracted to interesting combinations of letters detected parafoveally. Once the plausibility of the different control hypotheses has been established the task will be to identify the circumstances under which they can be seen to operate.

9. Moment-to-moment control of fixation location?

The simplest model of eye guidance effectively says that there is no moment-to-moment control of the locations of fixations during reading. This model can be rejected on the basis of any demonstration of saccade length being sensitive to the words being inspected. Demonstrations of this sensitivity are readily available. The evidence in favour of visual control and of cognitive control is necessarily evidence against this view of pre-programmed movements. Alternative theories argue for control by the visual characteristics of text ahead of fixation, or for control following the parafoveal pre-processing of syntactic and semantic characteristics. If we tend to look at the most interesting or informative parts of a text, then the question arises of how we knew they would be worth looking at before looking at them. Did we fixate these parts by chance, or was there some processing prior to fixation? If there was some pre-processing, then what kind of information can be extracted from parafoveal vision and used to guide our eyes?

If a reader's eyes halt at random as they progress from left to right along a line of text, then there should be no effects upon eye movements of textual features or word features. Any evidence of the influence of word frequency, length, or syntactic class would be evidence against this position. A simpler demonstration of the sensitivity of the guidance system to the characteristics of the text comes from a procedure used by Hochberg (1970), and which was used in support of his 'peripheral search guidance' model. By this model future fixations are guided by the presence of informative physical features. When an informative feature is detected a saccade is initiated so that the foveal inspection can determine the identity of the critical pattern. Hochberg's procedure involved the corruption of text by filling in the spaces between the words. Compare the ease of reading:

@In@her@urgent@anger@she@slapped@his@flabby@face@a@dozen@times@.
with:
Despite the surface good humour he must have been mildly upset.

Although the first sentence is readable, the absence of word spaces does disrupt the easy transition from left to right, and Hochberg found that the disruption was greater for good readers than for poor readers, in a sample of children. If the guidance mechanisms relies upon peripherally detected information about word shapes and word boundaries, then the reduction of this information should disturb the smooth flow of rightwards fixations. Evidence in favour of this suggestion has been provided by Rayner (1975), who found that word shape was influential from as far as 12 character spaces from fixation. The evidence from the studies of the effects of text corruption can also be interpreted in support of the Visual Control Hypothesis, but the data are not unequivocal. An alternative interpretation, in terms of the Process Monitoring variant of the Minimal Control Hypothesis, would suggest that saccades are shorter with corrupted text because of an early decision to move in smaller steps along a line. The difficulty in recognising words is increased by adding the processing required in identifying the locations of the words, and this increase in the cognitive loading might be used to determine the constant saccade length for the remainder of the text. Eye-movements would then be sensitive to the difficulty of recognition, but not on the moment-to-moment basis suggested by the Visual Control Hypothesis.

10. The plausibility of the Linguistic Control Hypothesis

Text-averaged saccade lengths cannot be used to distinguish between the Visual Control and Minimal Control Hypotheses. Better evidence of the moment-to-moment control of eye guidance comes from the observation of fixations upon critical words within texts. This evidence is necessarily harmful to the Minimal Control Hypothesis, and if it shows effects of the linguistic nature of the text rather than effects of purely visual features such as word length and word shape, then it establishes the plausibility of the Linguistic Control Hypothesis. The evidence to be considered here claims to demonstrate selective patterns of fixations according to grammatical and contextual constrains, according to associative relationships between words within a text, and according to the linguistic information available within a word.

Rayner (1977) asked the question of whether the locations of fixations can be determined by syntactic variations in sentences. His subjects read short passages which included sentences such as: "The policemen took the criminals out to lunch" that is, the critical sentences all had the same "The *subject verb* the *object prepositional phrase*" structure. Considering only forward fixations, because the guidance hypotheses are concerned with how we know how to inspect an unseen text, Rayner found that content words were fixated more often than the definite article and, more interestingly, that the saccades upon leaving the subject words were shorter than those recorded upon leaving verbs and objects. Rayner suggested that this resulted from the close proximity of subjects and verbs, with a fixation upon a subject being followed immediately by a fixation upon its accompanying verb. While demonstrating the effect of linguistic constraints upon eye-movements, Rayner's experiment is not without its problems. The failure to control for effects of word frequency and the omission of an analysis of the prepositional phrases make for a less than straightforward interpretation (see Underwood, 1985).

A number of lexical characteristics have been shown to influence fixation location, and the length of a word can be shown to affect the length of the saccade which results in its fixation. O'Regan (1979) demonstrated that the eye lands further into long words than it does when encountering short words, indicating that information about word length is both available from parafoveal vision and can influence fixation behaviour. Furthermore, readers make longer saccades when leaving long words than when leaving short words. Similar effects were reported for words of high redundancy. O'Regan found that the word "the" was often not fixated at all, in comparison with other three-letter words, a result similar to that reported by Rayner (1977). Interestingly, O'Regan's contrast between "the" and auxiliary verbs such as "was" and "had" was smaller than that between "the" and content verbs such as "run" and "saw". This difference may reflect the higher frequency auxiliaries being more recognisable with the limited visual information available in parafoveal vision. This result has been replicated by Carpenter and Just (1983), who found that "the" was fixated 40% of the time, "and" was fixated only 29% of the time, and that three-letter function words ("was/can/off/for") were fixated less often (47% vs 57%) than three-letter content words ("act/red/use/not"). O'Regan's (1979) finding of a difference in the sensitivity to main verbs and auxiliary verbs, when each is contrasted with responses to the definite article, has at least two interpretations. It may be an indication that a word's frequency can determine the probability of fixating that word, or it may indicate that the syntactic class of an as yet unfixated word is processed sufficiently early to influence the location of the next fixation.

11. Guidance of textual constraint

The importance of textual predictability during the reading of passages was demonstrated in an interesting study of fixation probabilities reported by Ehrlich and Rayner (1981). Contextual facilitation was observed by presenting target words in a passage which was highly predictive of the target or, for other subjects, in a contextually neutral passage. When the targets were predictable they were fixated less often than when they were in the neutral context (51% vs 62%), and for targets which were fixated, predictable words received shorter fixations than did the same words in the neutral contexts (221 msec vs 254 msec). Here is evidence of the influence of context upon the amount of processing given to a word, with unpredicted words receiving more attention than those which are constrained by the context of the passage. From our earlier discussion of how it is that we know where to look next it is interesting to note that this experiment provides further evidence of the use of parafoveal vision. Why should unconstrained words be fixated more often than constrained words if their significance has not been detected prior to their fixation? We must conclude that during a fixation prior to the target an area of interest or high information can be detected, and that this results in the selection of this area for a future fixation.

Why should a word be recognised more easily when it is predicted by the passage rather than when the passage gives no suggestion of its appearance? According to the logogen model of word recognition, evidence is collected from a number of sources prior to a threshold being exceeded (Morton, 1969). When no context is available, a clear presentation will provide evidence faster than a degraded visual presentation, but if the word has been preceded by a constraining context, then evidence will have been collected prior to the visual presentation of the word. In this case the reader may rely less upon visual information than when the context is not available or when the context makes prediction difficult. The Ehrlich and Rayner experiment also investigated the role of visual information when contextual constraints were varied. If readers are making less use of the visual features of the words, then they should be less able to detect spelling errors, and the eye movement data confirmed this prediction. Target words were sometimes replaced by words which shared all but one letter and which were totally inappropriate in the sentence. For example, in the sentence:

"The coast guard had warned that someone had seen a shark off the north shore of the island".

the word "shark" was the target, and was sometimes replaced by the word "sharp". Misspellings were fixated more often than their counterparts, although a second experiment indicated that this only holds true for misspellings which change the overall shape of the word. Misspellings also received longer fixations (240 msec vs 215 msec), but the critical data came from questions asked about misspellings at the end of the experiment. Readers would sometimes fixate a word but not report that it was misspelled. This occurred with 13% of the misspellings, and 88% of these were words in the constrained context passages. The context had induced less reliance on the visual features of the target, in line with the predictions of Morton's (1969) logogen model, and is a problem only too familiar to anyone who has proofread passages for spelling errors.

12. Guidance by lexical constraint

The effects of word associations upon eye movements were investigated by Kennedy (1978) by having adults read a three-sentence passage. A word in the final sentence was sometimes preceded by a word in the first sentence which was a good associate, and sometimes the priming word was replaced by a non-associate which fitted into the sentence. The results confirmed an influence of previously read material upon eye movements, but in some ways were rather surprising. Readers' eyes arrived at the second word faster when it was primed than when it was unprimed, suggesting perhaps that a primed word can attract fixations.

Kennedy's second result was that readers spent longer looking at a primed word than at an unprimed word. If the effects of priming are to aid recognition, then why should priming result in longer fixation durations? This result is difficult to reconcile with either the fast-automatic spreading activation mechanisms or the slower conscious attentional mechanism of contextual facilitation (cf. Posner and Snyder, 1975), but may have been due to the primed word engaging the reader's attention while lexical cross-referencing takes place. Having just read a sentence containing the word "hill", the appearance of the word "mountain" may result in indecision as to whether the words refer to the same object. This indecision would lead to a longer fixation duration while the integration calculation was completed.

A slightly modified version of Kennedy's experiment lends support to this integration interpretation. Kennedy used more priming trials than non-priming trials, and this may have encouraged subjects to search for the associates as part of what they perceived as the task. In a repeat of the experiment, Kerr and Underwood (1985) reduced the ratio of primed to unprimed sentences and found a result exactly opposite: readers spent 40 msec longer on the first fixation on an unprimed word than they did on a primed word. Using the same experimental materials, we also found priming effects of the same magnitude in two tachistoscope experiments, one a single-word lexical decision task and the other a single-word naming task. The facilitation effects are comparable, but of course not necessarily of the same origin. One difficulty in deciding whether the effects have the same origin is in not having a neutral condition against which to compare the primed and unprimed words: when sentences are used they can only be compared against each other. Lexical decision tasks can present non-words as the priming stimuli, and compare their effect with the effects of associated and unassociated words. Our experiment with sentences could only report that associated word-pairs resulted in shorter fixation duration upon the primed word than did unassociated pairs, but this does not say whether the primed words enjoyed facilitation or whether the unprimed words were impeded by the necessity of a more difficult integration calculation.

These experiments establish the case for a model which argues that, at the very least, visual features in the parafovea of vision are analysed to the extent that they can be used to guide the location of the next fixation.

13. Reading with parafoveal vision

Skilled readers are influenced by the information ahead of the current fixation, but the extent of processing is still a matter of investigation. Information about word length and word shape can be shown to influence the eye guidance mechanism, and other experiments demonstrate semantic processing of non-fixated words. This raises the question of whether identification of the physical features of words extends to identification of their lexical and semantic features.

Unattended, parafoveal words can influence behaviour in experiments using displays too brief to allow eye movements, and in which simple naming or decision responses are required. The nature of the influence depends upon the relationship between the immediate experiences of the reader and the parafoveal word, and the effects are observable when an associative relationship exists between what the reader is looking at or has recently seen and the words which occupy a location which would be a saccade to the right if an eye movement was possible (e.g. Underwood, 1976, 1981; Underwood and Thwaites, 1982). Although our tachistoscope experiments establish that parafoveal words are processed for meaning, they have used static displays, and bear little resemblance to normal reading. They do demonstrate that the meaning of parafoveal information is recognised however, and this lends plausibility to the notion that information about unfixated words can be used when reading. The case for parafoveal pre-processing is further supported by two studies reported by Balota and Rayner (1983), in which readers moved their eyes to name a previously parafoveal word, and by Jennings and Underwood (1984), in which readers named sequences of words, some of which had previously been presented in parafoveal vision. In both experiments reading performance was facilitated by the presence of associated words in the parafovea of vision.

Balota and Rayner (1983) briefly displayed a word which was accompanied by a non-word a few degrees to the left or to the right. The reader's task was to fixate the parafoveal stimulus as quickly as possible and then name it. During the saccade the non-word was replaced with a word. The important feature of this experiment is how the new word varied in its relationship to the fixated word and to the previously exposed parafoveal word. For the fixated word "reptile" the (old) parafoveal non-word was "snckks", and this could be changed to "snakes" (associate of the fixated word and visually similar to the parafoveal non-word), or to "lizard" (associate only), or to "sneaks" (visually similar only), or to "limits" (neither associated nor visually similar). If parafoveal information can be used to facilitate a response when that information is alter fixated, then the speed of word naming should vary according to the similarity between the parafoveal non-word and the named word which replaced it, and the answer was that it did.

Readers named the target faster when it was an associate of the previously fixated word, and when it was visually similar to the previously presented parafoveal non-word.

The second demonstration in support of parafoveal processing also looked for effects in sequences of fixations. Jennings and Underwood (1984) observed influences of previously parafoveal information upon word naming, in a tachistoscopic simulation of a sequence of eye movements. The word presented on one trial (trial N) was sometimes an associate of a word presented parafoveally on the next trial (trial N + 1). The word on trial N might be expected to aid the recognition of the parafoveal word on trial N +

1, but report was never requested. However, the naming of fixated word on trial N + 1 was affected by the presence of this association. As in Underwood's (1976) investigation of the influence of parafoveal words, this was an inhibition effect and this demonstrates that the immediately previous experiences of the subject can interact with currently available parafoveal information. The sequence of presentations did not stop at that point, and on trial N + 2 the previously parafoveal item was sometimes presented as the fixated item. So, this word was an associate of the fixated item on trial N, and was the same word as was sometimes presented in the parafovea on trial N + 1. In this case a facilitation effect was observed, with a faster naming response being seen for a word previously presented as a parafoveal item. This experiment can be viewed as a frame-by-frame version of Kennedy's (1978) priming experiment, and with a similar result. The presence of an associate in one sentence (equivalent to trial N which had a word with a critical parafoveal word), decreased the time taken to fixate upon a target in the third sentence (equivalent to trial N + 2). This result is the equivalent of our associate facilitation effect.

The parafoveal processing hypothesis suggested by these experiments does not require that the word to the right of fixation is fully identified prior to the fixation. If it is detected as being associatively related to the word or idea currently being considered then it might be marked as being a word in need of further scrutiny. Similarly, a word with an unusual orthography, or a word which is not readily recognised through parafoveal processing, might also be marked as requiring further attention. Parafoveal processing does not have to be comprehensive processing, but provides a means by which our eyes are guided to those parts of the text which are most useful. Neither does each parafoveal display need to be analysed, as a reader can proceed through a sentence quite skilfully analysing only the words fixated directly. Parafoveal processing would ease the integration of words, and provide some fine tuning for the guidance mechanism, but it cannot be regarded as being essential for skilled reading any more than it needs to provide a complete analysis of the words ahead of fixation.

14. Fixation locations within words

Words which are too long to be apprehended in a single fixation present readers with a special problem: how do we know where to look within these words? The Minimal Control Hypothesis suggests one applicable strategy for the reading of sentences, that is, to move the eyes rightwards by a reasonably constant distance at the end of each fixation. This strategy would not make use of any information gained from parafoveal vision or from the currently fixated part of the word, and only previously fixated material would determine fixation patterns. Long words provide a varied field of data for the lexical system to process, with different morphemes in these words having independent frequencies in the language as a whole. A long word may have two or three morphemes, with a combination of common and uncommon components. For example, in a word such as "supervisor" the first few letters are shared with a large number of other words, and in this sense they provide relatively less information about the word than the first five letters of "moralistic". We have seen from the studies of fixation patterns upon the words of a sentence, that fewer fixations are given to the less-informative contextually predictable words (Ehrlich and Rayner, 1981). These results lead to the question of whether, if predictable words attract fewer fixations, then do predictable parts of words influence the guidance mechanism in a similar way?

There are fewer fixations on predictable words, and this results from an interaction between the reader's linguistic expectations and from the early processing of words ahead of fixation. Strong expectations result in less of a need to rely upon the visual information contained in the expected word. Upon reading certain words, other words must follow if the text is to remain coherent - an adjective requires a noun, for instance - and when our expectations are violated then our fixation patterns change, as with a deleted agent in a sentence such as "The criminal was arrested by the river." When our linguistic knowledge allows us to predict the range of words which will appear in a specific slot in a sentence, then we should require less visual information from the page. One of the results from the Ehrlich and Rayner (1981) experiment on the detection of misspellings was that a strong linguistic context induces less reliance upon the visual information available in a word. On a smaller scale the same relationship holds within words. We tend to give more attention to the parts of words which are predictive of the identity of the word. This attention takes the form of more fixations upon informative sequences of letters, and longer fixations when they are made.

Support for this notion of varying visual attention to the parts of words according to their informative value comes from investigations of the inspection of long words. O'Regan (1984) presented his readers with a single long word which was presented asymmetrically to fixation. The first fixation on the word was therefore either towards the beginning or towards the end, and the words themselves were either informative towards the beginning or the end. Sensitivity to the location of the information was shown when the first fixation was at the beginning of a word such as "protagonist" (O'Regan used French speaking subjects), which is predictable on the basis of its first few letters in the way that "moralistic" or "breeziness" or "yearningly" are predictable for English readers shown the first few letters. When the initial inspection was upon the third letter from the beginning, then there was a longer fixation than if the word had been the ending-informative "extravagance" (cf., "multiplier" or "concertina" or "circumvent" for English readers). The first fixation was sometimes imposed upon the less-informative end of the word (the third letter from the end). With such an initial fixation, saccades leading to the second fixation tended to be longer and therefore resulted in an inspection of the more-informative end, than when the first fixation provided identifying information. This experiment provides good evidence of the moment-to-moment control of both the duration of the first fixation and the length of the first saccade within the word. It is also possible to show that the inspection of a previously unfixated word is sensitive to the distribution of information, and this demonstration has implications for the eye-mind assumption.

In a series of experiments using English and Finnish readers we have confirmed the effects reported by O'Regan (1984) using long words presented asymmetrically around the point of fixation (Underwood, Clews and Wilkinson, 1989; Hyona, Niemi and Underwood, 1989). In further experiments the words were presented in short sentences, which the subjects read prior to a comprehension test. In these experiments the initial fixation on the critical word was not determined by the display but by the reader (Hyona et al., 1989 Underwood, Clews and Everatt, 1990; Everatt and Underwood, 1992). The critical word was part of a sentence, and as it was never the first word in that sentence it was approached by the reader's eyes from part of the sentence to the left of the word. The results of these experiments suggest that longer inspections are given to the first half of a word if the information is at the beginning than if the information is at the end. Conversely, the inspections in the second half are longer if that is where the information is. A second result concerns the probability of fixation: there are more fixations in the first half for informative beginnings and more fixations in the second half when the information is at the end rather than the beginning.

The final result from these studies indicates that the sensitivity starts before the word has gained foveal inspection: the first fixation upon the word is nearer to the end of the word when the information is at the end. Although this effect is small in terms of the difference between the two fixation locations, it is obtained in a number of separate experiments and in both the English and Finnish studies. Full confidence cannot be given to this result however. We have obtained it in most of the experiments set up to observe it (Underwood, Clews and Everatt, 1990; Everatt and Underwood, 1990), but not in all of them (Underwood, Bloomfield and Clews, 1988).

This effect, if good, would demonstrate that fixations can be influenced by the information ahead of fixation, but it is not clear whether this is orthographic or lexical information. It may be the case that highly familiar orthographic patterns within words can result in saccades which will avoid their fixation - this may be what happened when O'Regan (1979) found the word "the" being avoided. Alternatively, it may be that lexical information about the component morphemes can be used by the eye guidance mechanism. Whether the useful information is orthographic or lexical, however, non-fixated information is able to affect the location of the next fixation, suggesting that parafoveal information is used by the mechanism which decides where we should look next, in general agreement with the predictions of the Linguistic Control Hypothesis. Further evidence provided by Carpenter and Just (1983, p. 283) agrees with the parafoveal processing hypothesis. They found that occasional fixations upon function words were followed by the non-fixation of the content word to the right. Further, the amount of attention given to the function word was affected by the frequency and length of the skipped content word. In this case, the gaze duration on the function word was reflecting the processing of the content word available in the parafovea.

15. Psycholinguistic Processes are Illuminated by Watching Readers' Eyes

Although we have to show caution when making conclusions about processes from eye movement data, due to the possible lack of viability of the eye-mind assumption, this does not necessarily mean that we should abandon eye movement data altogether. A great deal of information can be gained from the study of eye movement behaviour, and the following is an illustration of the power of eye movement measures used to aid our understanding of what is going through the reader's mind when texts are inspected. The discussion will look at three main features of visual linguistic processing, the identification of individual words, the processing of syntax, and the influences of semantics.

The evidence discussed so far suggests that information outside of the centre of fixation is processed to some extent, possibly to guide saccadic movements, but there is still evidence that the processing demands within the current fixation on a word are still strongly influenced by properties of a fixated word, and less so by information around it. In one example we have mentioned, the data of Carpenter and Just (1983) suggest that the frequency of the words around the fixated word do not affect fixation durations on that word. Along with this the evidence of Rayner and Pollatsek (1981), that there is little effect on current fixation durations when we eliminate word boundaries by replacing them with Xs, this also suggests that current fixation durations show influences of the processing of the current word, and less so information around it (although this may not be the case when words are skipped: cf, Hogaboam, 1983). Also, even if the eye-mind assumption is doubted, and we assume that information about lexical access is processed in the parafovea, we can still make conclusions about this processing by carefully manipulating items in foveal and parafoveal vision, as we shall see.

16. The identification of individual words

If we accept that fixation durations are strongly influenced by the processing of the fixated word, what factors can be shown to influence these fixation durations? And what can we conclude about linguistic processes from these influences? One factor that does not seem to influence processes within fixations is syllable number. Crowder (1982) presents evidence that fixation times are not affected by the number of syllables in a word, suggesting that initial processing on a word is not affected by the ease with which a word can be processed as a phonological unit. This suggests that articulation does not play a role in the initial processes which are applied to a word, and if sound codes are important in visual language processing, we must look for another role for them. As we have discussed already, the durations of fixations on a word are affected by the frequency of occurrence of that word within its home language. Inhoff and Rayner (1986) found that word frequency affected the duration of the initial fixation within a word when word length was controlled. This was of the magnitude of 30 or so milliseconds between high frequency and low frequency words. Since the initial fixation is sensitive to this effect it suggests that the initial processing of the word (identification perhaps) is open to frequency influences.

Related to this are the findings of Lima and Inhoff (1985). Their evidence suggests that the number of lexical neighbours possessed by a word also affects the durations of fixations on that word. Fixation times were longer on words such as "clown", which have a large number of lexical neighbours possessing the same initial letters ("clo"), compared to words such as "dwarf", which have very few neighbours with the same initial letters ("dwa"). This suggests that distinctive letter combinations within words can speed up the processing of a written stimulus. Since the effect was more reliable on initial fixations than gaze durations, it again suggests that the effect was manifest more within earlier processing of the word. The evidence from our own studies on informative areas within words also suggests that the reading system is sensitive to such factors early in the processing of the stimulus.

The length of the word also affects eye fixations, with both duration and number of fixations being increased when longer words are viewed (cf., Blanchard, Pollatsek and Rayner, 1989; Rayner and Pollatsek, 1987). If a word's length affects the fixations upon it, this suggests the possibility that words are decoded letter-by-letter. Thus a word would be recognised by identifying individual letters and combining these to form the appropriate word. Increasing the number of letters in a word means that the number of letters that have to be recognised has increased, and so the amount of processing necessary on a word has also increased. Evidence from studies by McConkie, Zola, Blanchard and Wolverton (1982) and Blanchard, McConkie, Zola and Wolverton (1984) however, suggests that this may not be the case. The McConkie et al. study changed words that subjects were reading back and forth during saccadic movements around those words, when visual sensitivity is poor. Fixation behaviour was not detrimentally affected by these changes. Words were changed from, for example, "bears" to "peaks" and back again, and subjects were asked to indicate which word they had seen. If initial letters were processed before later letters, then, in certain cases "breaks" would be reported. This was not the case: fusions of the two stimuli were not reported. In the Blanchard et al. study, a word was changed, at certain times relative to fixations around it, to another word that possessed a different initial letter or a different fourth letter. Again subjects had to indicate which word had been presented. Subjects produced one or the other word, or both words. If words were processed letter-by-letter we would expect that the first word presented would be much more likely to be given when the initial letter is changed than if the fourth letter is changed, because the initial letter is more likely to have been processed before the change than the fourth letter. This was again not the case. Reports of the first presented word were about as likely for initial letter changes as fourth letter changes; in fact slightly higher for fourth letter changes. This evidence suggests that whole word information (or no word information) is carried between saccades, and that the initial letter is not more likely to be processed sooner within a fixation around that word than the fourth letter. Letter information would thus seem to be processed in parallel to some extent.

On the other hand Inhoff (1987) found that giving subjects part-word information in the parafovea did aid recognition of the whole word, but that this was greater when parafoveal preview was of the first morpheme of a compound word. For example, parafoveal preview of "cow" in "cowboy" led to greater reductions in foveal fixations than parafoveal preview of "car" in "carpet". This suggests that individual morphemes may be decoded to some extent in the parafovea. It also gives weight to views which consider that morphemic decomposition occurs within reading. These possibilities were tested in a study by Lima (1987). Prefixed ("remind") and pseudoprefixed ("relish") words were placed in sentences, and subjects' eye movements were recorded while they read for comprehension. Previews of the targets were such that either the whole word was present before fixation upon it, or only the prefix was present before fixation upon the word, or none of the word was present before fixation upon it. Preview advantages (shorter fixation times on the targets) were found for whole word previews, but not for prefix previews, and there was no interaction with type of word. However, initial fixations upon the target words were shorter for prefixed than for pseudoprefixed words. This suggests a disadvantage for decomposing a pseudoprefixed word compared with a true prefixed word, as the decomposition views of Taft and Forster (1975) would expect. However, these findings suggest that this decomposition process, if it occurs, occurs on fixations within the word, not parafoveally. Since the initial fixation seems prone to these effects (gaze duration effects were less reliable), the results are consistent with this effect being produced by initial processes upon a word; possibly, decomposition prior to lexical entry, or at least affecting lexical entry.

The evidence presented here suggests that eye movement data is affected by word length, frequency of occurrence within a language, and the distinctiveness of information within that word, but not by the number of syllables. It also suggests that letters within words are processed to a large extent in parallel, except in the case of morphemically complex words, where there is evidence to suggest that these words are decomposed at some stage during their processing. The eye movement data thus suggests a word recognition system that is sensitive to visual information rather than verbal information, a system whose processing of a word is aided by distinctive features within the word and by the frequency of use of that word within the language, and that to a large extent processes letters within a word in parallel.

Once a lexical entry has been accessed we have to place it within a discourse framework. This involves accessing a meaning, placing a syntactic role on the item, and integrating that meaning with information elsewhere in the text. First of all then a meaning has to be accessed. A problem arises here because a great many words have multiple meanings. How do we decide which meaning is appropriate in the given discourse? The obvious solution is to use contextual information to infer the implied meaning. However, what happens when there is no such information? Do we make a guess at the implied meaning, use some heuristic to decide which meaning is most likely to be implied, or do we wait until information later in the text makes the implied meaning obvious? Eye movement data regarding such questions have been very informative as to the functioning of the linguistic processor, and its processing of syntactic material. We will discuss these findings next.

17. The processing of syntax

The studies of eye movement behaviour we shall discuss here mainly concentrate on "garden pathing" effects; that is, leading the reader to make one interpretation of the text and then giving them contradictory evidence. The usefulness of these studies is that they reveal recovery processes which are applied by the reader after a syntactic processing error has been made. For example, Carpenter and Daneman (1981) gave subjects sentences in which homographs (like "tears") appeared, and provided the readers with initial information that implied the homograph meant one interpretation (water running from the eye), then later information that it meant a different interpretation. For example, they gave the subjects sentences like:

> "Cinderella was sad because she couldn't go to the dance
> that night. There were big tears in her brown dress".

This produced long fixations on the word "dress" and more regressions back through the text. This evidence suggests that contextual information is used to disambiguate information within a text. The several meanings of "tears" may or may not be accessed, but even if they are, only one of the meanings is interpreted from the discourse. The evidence also suggests that individuals will interpret information in one way until information for a different interpretation is encountered. This incompatible information then leads to a slowing down in processing of the text, and, on occasions, the necessity for re-readings of the text.

Are all the meanings of a word accessed? And does this occur even if there is prior contextual information indicating a single interpretation? Experiments using words with multiple meanings performed by Rayner and Duffy (1986) and Duffy, Morris and Rayner (1988) provide evidence that the different meanings are actually accessed even if prior context implies only one of the meanings, but that one meaning is quickly interpreted from the accompanying contexts. The Rayner and Duffy experiment gave subjects sentences containing ambiguous words whose meanings varied as to how dominant one meaning was within the language. For example, "boxer" is more usually interpreted as a person who fights than a breed of dog. These were compared to ambiguous words whose meanings were equally dominant ("coach", American subjects were used). They found that gaze durations (first fixation durations were unreliable) on ambiguous words with equally dominant meanings were longer than those on control words (non-ambiguous words) whereas, gaze durations on ambiguous words with a dominant meaning did not differ from those on control words. They suggested that this was because both meanings of the former ambiguous words had to be accessed (increasing processing time), while only the dominant meaning of the latter was accessed. If prior context makes one of the meanings of the equally dominant ambiguous words more plausible the difference in fixation duration is removed. This implies that context is indeed used to disambiguate ambiguous information, as intuition would suggest. However, in the follow up study Duffy, Morris and Rayner presented ambiguous words with a prior context indicating the less dominant meaning. Here gaze durations increased compared to those sentences where there was no disambiguating prior context. This indicates that the more dominant meaning is accessed to some extent automatically and interferes with the interpretation of the less dominant meaning, even with prior disambiguating information. There was also a slowing down in reading speed when later disambiguating information indicated the less dominant meaning, and, even more interestingly, when an ambiguous word with no dominant meaning was followed by a context indicating one of the meanings. This suggests that subjects make one interpretation of the ambiguous information and stick to that until information to the contrary is encountered. When there is a dominant meaning, this meaning if selected, but also when there is no dominant meaning, one is still selected. In terms of the present discussion this again established that one meaning is chosen during the interpretation of a word within a sentence, rather than all possible analyses being undertaken, or some delay in decisions occurring.

Evidence from eye movement data of the processes involved in syntactic parsing comes from a study using similar garden pathing sentences by Frazier and Rayner (1982). They used sentences such as:

> "Since Jay always jogs a mile this seems like a short distance to him."

and these were compared to sentences such as:

> "Since Jay always jogs a mile seems like a short distance to him."

Here the phrase "a mile" is interpreted as the object of "jogs", due to a hypothesized characteristic of the linguistic processor, that of "early closure". Basically, early closure states that new information should be attached into the phrase currently being processed. New words are added to current phrases whenever the rules of syntax allow, rather than being allocated to new phrases. Sometimes this strategy is inappropriate, as in the second of Frazier and Rayner's sentences. This hypothesized characteristic of the syntactic processor predicts that the second sentence should be harder to read than the first version, since in the second version early closure does not hold and "a mile" is not the object of "jobs". Evidence for this viewpoint came from the findings of longer reading times for the second version. There were also longer initial fixations in the region of the sentence which disambiguate the sentence ("seems like"), and more regressions to the ambiguous information ("jogs a mile"). This eye movement data suggests that the initial, incorrect, interpretation of "a mile" is processed, and the following information ("seems like") indicates such an inappropriate interpretation of the text, as in the initial interpretation of "tears" in the Carpenter and Daneman (1981) study. This again leads to longer fixation durations around the disambiguating information, as the detection of the anomaly is made, and the reinterpretation attempted. It again also leads, on occasions, to a regression being made to the ambiguous information to aid in the reinterpretation. Thus, interpretations appear to be immediate and specific, and reinspections of sentences influenced to a large degree by the information within the sentences. As in the Carpenter and Daneman study the evidence suggests that immediately upon encountering disambiguating information, or information inconsistent with the present interpretation, regressive saccades occur to the ambiguous information.

In a recent study we have confirmed the general conclusions of the Frazier and Rayner study (Underwood, Briscoe and MacCleary, 1992), and extended the results to observe the effects of introducing a comma. The early closure sentences, which previously caused parsing difficulties, became unambiguous when the punctuation was introduced to indicate the phrase boundary, as in the following:

> "While Pam was washing, the baby started to cry"

compared with:

> "Though Ben read the book bothered him"

This syntactic significance of punctuation is beyond the scope of the present discussion, and suffice it to say that the comma can be regarded as non-syntactic information, and our demonstration of its effect upon parsing stands as a demonstration of an interaction of information from syntactic and non-syntactic sources. While some linguists consider the syntactic processor to be independent of other sources of textual information, the influence of the comma suggests that non-syntactic information can influence the parser. The main reason for mentioning this study is to introduce one reader who had a very atypical inspection strategy, and who will be discussed as the speed reader (HA) in the section of this chapter concerned with Individual Differences.

The evidence presented so far is consistent with the immediacy hypothesis of processing: that all the processing necessary for a word to be fitted into the meaning of a discourse occurs while the eyes are fixating that word. However, evidence from both the Carpenter and Daneman study and the Frazier and Rayner study suggests that this is not always the case. On occasions the fixation data show that problems with processing the disambiguating information occur after this information is encountered and the eyes have moved on to following information; this is shown by subsequent longer fixations, and regressions from the ends of sentences to ambiguity. This evidence suggests that full interpretations of a discourse may be delayed until later in the sentence, or even to the end of a sentence, or that reinterpretations are delayed in order to wait and see if further information aids in the processing of the discourse.

Frazier and Rayner's (1982) study also provides evidence for a second hypothesized characteristic of a syntactic processor, that of minimal attachment. Minimal attachment basically considers that we attach incoming material into the phrase marker currently being constructed. Frazier and Rayner's data are consistent with the views of minimal attachment, with sentences such as:

"The city council argued the mayor's position forcefully"

being compared to, and showing shorter fixation durations after "the mayor's position" than:

"The city council argued the mayor's position was incorrect"

Minimal attachment would consider that the "mayor's position" should be interpreted as the direct object of the verb "argue", as in the first sentence, but not as in the second. Frazier and Rayner's data suggests that a syntactic processor is used to interpret a piece of text, and assign syntactic roles for information within a discourse. They consider that the operations of this processor are independent from those of a semantic, thematic, processor. However, there is evidence which suggests that linguistic processing may be heavily associated with semantics, and has given weight to theories viewing the reading processor as a much more interactive processor (see McClelland, 1987), or to views considering that separate syntactic processing is not necessary (cf., Schank, 1972).

Evidence against the independent syntactic processor viewpoint and for considering the linguistic processor as much more interactive is provided by a study by Taraban and McClelland (1988) which suggests that violations of thematic expectations slow down the reading of words after that violation. However, this study used a self-paced word-by-word reading technique by which a button-press is made to bring up the display of the next word in the sentence, and this may have produced unusual reading strategies from the subjects. Just, Carpenter and Wooley (1982) and Ferreira and Henderson (1990) discuss these problems and present evidence that such word-by-word procedures slow down reading and so may lead to abnormal strategies, such as the greater use of semantic information. The argument here is that of the value of an eye-mind assumption over a button-press/mind assumption. Although the eye-mind assumption may be open to doubt, as we have argued here, button-pressing procedures are probably even more problematic. A discussion of the values of different procedures such as these is beyond the range of the present discussion though, so let us move on to consider what evidence, if any, eye movement behaviour can provide in this area.

In a study of the influence of types of linguistic information upon eye movement behaviour Rayner, Carlson and Frazier (1983) found evidence which suggests that syntax and semantics are processed separately. They gave subjects sentences in which pragmatic knowledge could be used to parse garden path sentences appropriately. For example, they found that sentences such as:

"The performer sent the flowers was very pleased"

showed equally long fixation durations on the disambiguating part of the sentence ("was very") as sentences such as:

"The florist sent the flowers was very pleased"

There was no difference in syntactic parsing difficulties even though semantics would suggest that florists are more likely to send flowers, leading to the inappropriate interpretation, whereas performers are more likely to be sent flowers, leading to the appropriate interpretation. In both cases it appears that subjects interpreted the sentence as suggesting that the performer/florist sent flowers rather than received flowers, as the hypothesized characteristic of late attachment would predict. The evidence here suggests that pragmatic knowledge about the actors in this discourse did not influence the syntactic interpretation placed on the sentences. Similar evidence for the separation of syntactic processing and pragmatic knowledge has been found by Ferreira and Clifton (1986) who found little difference in the reading of sentences like:

"The defendant examined by the lawyer turned out to be unreliable"

and:

"The evidence examined by the lawyer turned out to be unreliable"

This was the case even though in the second sentence semantic knowledge should tell us that "evidence" is more likely to be examined than to examine something else.

How distinct are the syntactic and semantic processors? Are they closely interactive or are they the same processor showing differing effects on syntax and semantics? The evidence is not entirely conclusive. In certain cases they seem to be quite distinct processors, as the Rayner, Carlson and Frazier

(1983) and Ferreira and Clifton (1986) studies suggest. However, in other cases they seem to be more closely interactive, as the Taraban and McClelland (1988) and Underwood, Briscoe and MacCleary (1992) studies suggest.

18. Semantics, pragmatics, and inferences

The studies discussed so far suggest that to a large extent information is processed as it is fixated. It suggests that readers interpret the information to a large extent before leaving that word. They do not seem to retain a number of possible interpretations and move on waiting for clarifying information. The question then arises as to how far do individuals make interpretations about the information they are reading. Do they make complete and complex interpretations about the information they are reading within the discourse or just enough to fit that information with what has just been read. Basically, the question is to decide what sort of information goes into the processing of a visual stimulus?

Eye movement data has provided information about the type of inferences that readers make about a word when presented with it in a discourse. For example, Just and Carpenter (1978) found that fixations on the word "killer" were longer if it was preceded by the phrase " . . . the millionaire died" than if it was preceded by the phrase " . . . the millionaire was murdered". This they considered was because an extra inference about what happened to the millionaire had to be made (i.e. that he was killed) in the former condition, which did not in the latter. Similar data is presented by Kennedy (1978) who presented subjects with a series of sentences of the form:

> "It is unwise to wander on a mount. People can get lost there.
> A hill is not always easy to climb."

and compared this to:

> "It is unwise to wander on a track. People can get lost there.
> A hill is not always easy to climb."

Here longer reading times were recorded on "hill" when the preceding sentence referred to "mountain" rather than to "track". This can be interpreted as indicating that the reader has to spend time making a connection between the reference to hill and its antecedent, mountain. When track is the prior object such a connection need not be made and the sentences are treated somewhat separately. Thus readers process information and try to connect it with antecedent information. A certain amount of time is used for locating and connecting references to the same concept. The evidence here is that this again occurs during fixations on the word referred to previously. We shall return to this later.

O'Brian, Shank, Myers and Rayner (1988) found evidence that individuals make general-to-specific inferences about words within a sentence. For example, they gave subjects the following sentence:

> "All the mugger wanted was to steal the woman's money. But when she screamed, he stabbed her with his weapon in an attempt to quieten her. He threw the knife into the bushes, took her money and ran away."

Gaze durations on knife in the last sentence were the same if "weapon" was presented compared to when the same paragraph was used with the word "weapon" being replaced by "knife" itself. This suggests that knife can be inferred from the word weapon in the previous sentence. When, however, the word "stabbed" was replaced by "assaulted", then gaze durations were longer on knife if it was not explicitly mentioned in a prior sentence. Thus, knife was inferred from weapon only when stabbed was mentioned.

An experiment by Kerr and Underwood (1984) found evidence that pragmatic information (knowledge about the world) about sex roles affects processing of pronouns. Subjects were given sentences in which a pronoun referred to a previous role. In the situation where the role and pronoun did not conform to sexual stereotypes (i.e. "she" referred to "surgeon") there were longer initial fixations on the pronoun, compared to the situation where there were no constraining stereotypes (i.e. "she" referred to "student"). This suggests that inferences about the sex of individuals in a discourse made due to stereotypical models of the world, which, when violated lead to a re-evaluation of the discourse. This, in turn, will lead to a slowing down in the processing of the pronoun. It is interesting to note that fixation durations were slowed down on initial fixations, gaze durations and total fixation times. This suggests that the effects of violating such expectancies can be manifest in early as well as late processing of the stimulus.

Continuing with this theme, it also appears that semantic information such as gender can be used to disambiguate an interpretation of a discourse. Basically, there is evidence to suggest that the gender information of a pronoun can be used to distinguish the prior referent of that pronoun before information about causality can distinguish the same referent. Such evidence comes from a study by Vonk (1984). In this study subjects were presented with sentences such as:

1. "Alex lied to Andy because he smelled trouble"
2. "Alex lied to Anna because he smelled trouble"
3. "Alex lied to Andy but he smelled trouble"
4. "Alex lied to Anna but she smelled trouble"

Records of subjects' eye movements suggest longer fixations on the pronoun when gender cues are given (2 and 4), whereas there were longer fixations on the verb phrase following the pronoun ("smelled trouble"), and more regressions and refixations for second pass reading of the entire sentence, when no gender cues were provided (as in 1 and 3). This evidence suggests that pronoun antecedents were inferred earlier when gender information was available. This suggests two conclusions. Firstly, semantic information such as gender can be used early to make inferences about a piece of text. Second, and when no such overt semantic information is available, readers do not impose a referent upon a pronoun immediately upon encountering that pronoun. In the case where a pronoun can refer to either of two prior individuals, interpretation is delayed to some extent until later information is available to try to disambiguate the anaphoric reference. This appears to be a special case where full interpretation may be delayed until later in a text. Again, this is an example of a violation of the immediacy hypothesis.

What does this evidence tell us about the activity of the linguistic processor? Firstly, it seems that inferences about words within a discourse are made to quite a large extent. So readers appear to make inferences about gender identities of a subject in a discourse from their general scheme of the world. They appear to make inferences about actual examples of categories under specific circumstances: specific inferences about general statements. They also appear to make connections between different mentions of the same concept. The data thus suggest a complex processor that uses information within a discourse and within general knowledge to interpret a discourse in specific ways. It appears to make connections between information in a discourse, infer information from the discourse and from general knowledge, and process specific interpretations. These interpretations seem to be accomplished to a large extent within fixations on each word, and appear to be very specific. Only when information is encountered that contradicts the chosen interpretation is that interpretation questioned and others reinstated.

In the main, many interpretations are not carried through a discourse until one is unequivocal. There is also evidence for some of the rules that a linguistic processor may use to accomplish these complex tasks, such as using the dominant meanings of words within a language to interpret words with multiple meanings, and using syntactic parsing operations such as late closure and minimal attachment. These latter views have led to the viewpoint that the linguistic processor is made up of autonomous processors, each responsible for some operation within the understanding of a discourse.

Similar conclusions can be made from the available eye movement data for the role of semantics in the initial processing, or recognition, of a word. We have already discussed the evidence for semantic information aiding the recognition of a word in the parafovea in the study by Ehrlich and Rayner (1981). The study by Balota, Pollatsek and Rayner (1985) found, as did Ehrlich and Rayner, that words which appear in predictable context were more likely to be skipped than words that appeared in less predictable sentences. This evidence suggests that context and parafoveal information can combine to aid the recognition of an item by speeding up its access to a lexical entry. However, the evidence suggests that these effects may occur because of relationships between words within the context. Thus "cake" was more likely to be skipped if preceded by "wedding" than if "pies" was preceded by "wedding". These effects can thus be interpreted as occurring because of spreading activation through related items within a highly organised and connective lexical network.

A study by Carroll and Slowiaczek (1986) can be used to emphasise this point. They used category names to prime examples within sentences. Sentences contained high typical examples ("cotton"), or low typical examples ("canvas"), preceded by the category name ("cloth") or a neutral prime ("stuff"). Fixation times on the examples were on average shorter following category names than following neutral primes. This effect did not interact with typicality though. This suggests that context effects here were not predictive context effects, but were due to semantic relationships between words. This interpretation is backed up by a second experiment by Carroll and Slowiaczek in which highly associated items were placed into sentences. Here fixation durations on an item were shorter if it was preceded by a related item ("king"/"queen") compared to if it was preceded by an unrelated item ("ambassador"/"queen") or a neutral item ("man"/"queen"). There were no inhibition effects from unrelated items, which suggests that the effect was not a strategy effect (cf., Posner and Snyder, 1975). These effects then do not seem to be due to predictive semantic factors, but due to relationships between words. These relationships can be accounted for by the structure of the lexicon, and therefore we do not necessarily need to interpret top-down influences from higher semantic processes.

There is also another potential source for such semantic effects. The Balota et al (1985) study found that context facilitation effects on target words occurred not on the first fixation, but on the overall fixations on the target; and this spilled over to fixations after the target. Since identification processes would be assumed to occur early in the processing of a word, these effects suggest that later processing of an item, perhaps its integration into the context of the sentence, is also affected by semantic factors. Thus, as we have discussed above, integrating information together affects fixation data. This is usually later fixation data. A recent study by Duffy and Rayner (1990) suggests that effects of the distance between related information may be due to integration processes rather than semantic priming effects. The distance between a word and a prior referent or related word has been shown to affect fixation durations. Ehrlich and Rayner (1983), using related words, and Schustack, Ehrlich and Rayner (1987), using repetition of the same item, found that if the distance between these items is increased then fixation durations are also increased. However, Carroll and Slowiaczek (1986) failed to find such a distance effect. The Duffy and Rayner (1990) evidence suggests that such distance effects occur because new information is connected to prior information, and the further back in the text this prior information occurs, the longer such integration processes will take. Ehrlich and Rayner and Schustack et al. found distance effects because integration was required, whereas Carroll and Slowiaczek did not find distance effects because integration was not required. Their evidence also suggests effects on the fixation times after the target words, suggesting complete integration is not accomplished on the target word. Duffy and Rayner manipulated the typicality and distance of an antecedent. They found that if "weapon" was preceded by "sword" gaze durations were longer on "weapon" if "sword" was more distant. However, if "object" was preceded by "sword", there was no difference in gaze durations on "object" if "sword" was distant or near. They concluded that when a connection between items in a discourse is suggested then processing of that connection takes place, which takes longer if the prior information is further back in the text. Distance effects are thus due to anaphor resolution. The presence of "weapon" suggests a connection with "sword" and time is taken to make this connection, the presence of "object" is too general to suggest a connection and therefore time is not taken in searching for an antecedent. Schustack et al. also did not find distance effects in a naming task with the same sentences, again suggesting that such distance effects are not due to simple lexical access.

There thus appear to be at least two separate sources of semantic effects within reading. The first appears early in the processing of a word, and seems to be associated with semantic relationships between words within a discourse. This can be explained if it is assumed that lexical entries are connected to semantically related lexical entries, and that activation of one entry leads to activation in related entries.

There is no need to consider that information from higher level semantics feeds back to aid recognition processes. There are also semantic effects associated with integration processes. These occur within fixations within the word, and to some extent can spread over to fixation data after the to-be-integrated word. These integration processes involve finding prior references to the same concept, and seem to be fairly exhaustive. If prior references are further back in the text, time is taken to locate these references and make the integrative connections.

19. Watching psycholinguistic processes: some conclusions

Let us consider the three areas of interest here separately. Firstly, word identification processes seem to be sensitive to the length of a word, its frequency of use and whether it contains more distinctive letter combinations. The operations of these processes appear to be accomplished to some degree in parallel, with individual letters not being identified separately, although there is evidence that some form of morphemic decomposition may take place within these operations. These operations also appear to be fairly autonomous from other operations in the linguistic system. Identification does not appear to be influenced by top-down processes, but rather by processes within the lexicon itself. It also appears that the meanings of a word are accessed automatically to some extent, if not completely, and some later operation is used to distinguish the intended meaning of the word from other information in the text; although to some extent this process is affected by the dominance of a particular meaning. These processes seem to be accomplished to a large extent by fixations within the word in question, although there is evidence for identification processes occurring with prior fixations, and processes involved in determining the meaning of a word occurring with fixations after the word.

The eye movement data suggest that processes involved in syntactic processing are also autonomous from higher-level processes such as semantic and pragmatic recognition. Two characteristics have been distinguished in these data, that of minimal attachment and early closure. These characteristics can to a large extent explain the available eye movement data without the need for considering semantic/pragmatic information being used to aid in the production of a syntactic framework. Subsequent violations of this framework then will lead to a slowing down in the processing of the text, and the potential re-reading of that text. Eye movement data also suggest that such re-readings can be very specific to the source of difficulty. This suggests that to some extent structural information about the discourse is stored within the comprehension system.

Further processing of the text appears to be involved in making semantic connections between a number of mentions of the same concept. When a concept is referred to twice, processing time is taken up finding the prior referent, and connecting the concepts into the framework of the sentence. Another semantic factor that eye movement data provides information about is the inferences made by readers within a discourse. Here there appears to be evidence for readers making complex inferences about specific concepts from general concepts and evidence for subjects using stereotype pragmatic information about sex roles. There is also evidence for subjects using this knowledge at different times within the processing of a discourse. Complete semantic processing, anaphoric connection and inferencing, may not be accomplished within the fixation upon a particular word. Some evidence suggests that this may spread over into fixations after the word, and that this may depend upon the type of information available and the type of inference to be made. Thus, although in most cases a full interpretation of the word within a discourse appears to be made, there seem to be occasions when such interpretations are held over to see if further information is available to make the correct interpretation. There is even evidence that this may be held over to the end of the sentence.

20. Individual Differences in Reading

The evidence suggests that readers' eye movements are affected by various linguistic factors, such as ease with which a lexical item is accessed from memory, ease with which a discourse is syntactically parsed and the ease with which words are integrated into the semantics of the discourse.

However, this is not the complete picture when it comes to influences upon eye movements. Even if such variables are kept constant, there still appears to be a great deal of variance in individuals' eye movements. Both saccadic movements and fixation durations appear to be influenced by individual differences. For example, even if a fixed target is provided for subjects, there is still a great deal of variability in where the eye lands on these targets (Coeffe and O'Regan, 1987), and even if uncertainty about when and to where a saccade should occur is removed, fixation latencies are still highly variable (Salthouse and Ellis, 1980).

Why should this be? Why, if all the stimulus variables are controlled, should eye movements still display differing behaviour patterns, within and between individuals? One explanation is that such eye movement behaviour is random. However, as we have tried to argue above, this viewpoint does not seem compatible with the great deal of data showing that eye movements are influenced greatly by factors in the stimulus environment. One possible explanation for the variance in eye movement behaviour within individuals is that although messages to move the eyes are influenced by the stimulus, they are also influenced by internal processes which are used to programme saccades to new locations. Rayner and Pollatsek (1989) have used just such a process to explain why more than a single fixation occurs within a word. Another is that there is noise in the eye movement system that means that saccades may not be made to precisely the location desired. Individual differences in the reading system could also be used as an explanation of this variance between readers. Research into this area has found this may be related to reading experience. In a large scale study of individuals' eye movement patterns Taylor (1957) measured various components of eye movement data at various ages through school and college. These findings suggest that number and duration of fixations decrease with increasing reading age, as to do number of regressions back through a text.

If this is the case, do eye movements vary with reading ability? Evidence here is more mixed. Murray and Kennedy (1988) for example, found little evidence of differences between duration and reading ability when reading experience is controlled. Some research though has found evidence for a relationship between eye-movements and reading ability. For example, Olson, Kliegl and Davidson (1983) found that below average readers produced fixations which were 50 msec longer than normal readers. They also found a relationship between reading ability and regression backwards through a text. Kennedy (1987) considers that eye-movements may be able to discriminate between good and poor readers due to the type of regressions such readers show. He provides evidence that when the meaning of ambiguous words were later changed to a different meaning, and the regressions back to the ambiguous information measured, good readers were characterised by a single long saccade back to the ambiguous information, while poor readers showed numerous shorter regressions, or a return to the beginning of the sentence and re-reading of the sentence. This Kennedy proposed was due to better readers possessing better representations of the locations of words within a sentence.

In a study comparing reading comprehension and eye-movements in an adult subject population Underwood, Hubbard, and Wilkinson (1990) found a relationship between fixation duration and comprehension. Individuals who produced shorter fixations were also those who produced better scores on a comprehension test. However, they did not find a relationship between comprehension and number of regressions. A similar relationship was found between a comprehension test and fixation durations in Everatt and Underwood (1992). Here variance in performance on a gap comprehension test was predicted by a measure of vocabulary size and gaze durations. Shorter gaze durations were associated with better performance on the comprehension test, supporting the findings of Underwood, Hubbard and Wilkinson.

Taylor's (1957) findings also suggested an increase in the perceptual span of an individual (the area within which information can be accessed from the page) as reading experience increased. Here, the range of recognition (as Taylor put it, the number of words processed per fixation if a word is considered to consist of about ten characters) increased from about half a word when beginning to read to about one and a third words in adult life. Rayner (1986) on the other hand found that the perceptual span was only about 25% smaller for beginning readers than adult readers (11 characters compared to 15 characters). Again there is evidence that this is related to reading experience rather than reading ability. Underwood and Zola (1986) for example found no difference in the size of the perceptual span between good and poor readers when using children of the same age, suggesting that once development of a perceptual span is complete there is little variance in this that would account for individual reading skill. Also Everatt and Underwood (1992) found no relationship between ability to locate informative regions within words and reading comprehension scores in their adult population. This suggests that this ability is the same across the reading skills studied.

However, a larger span of perception could potentially be very useful in reading, and increasing span size is one of the methods used in the training of speed readers. These individuals seem to be able to read text very quickly, and, it is proposed, can do so without loss of comprehension. One of the claims put forward by proponents of speed reading is that parafoveal/peripheral vision can be used a great deal more than it actually is in normal reading behaviour, and so more information can be processed for fewer fixations and saccades. Speed readers have been reported to move their eyes down the left hand side, or middle, of the page to be read, fixating only once, if at all on each line (see Gibson and Levin, 1975). Average fixation durations have been found to be about, or slightly longer than, normal. This suggests that either a great deal of information is being picked up in the periphery of vision in these individuals, or they are guessing the text from reading one or two words per line.

21. The problem with speed readers is . . .

From the study of early closure sentences with and without punctuation we obtained a number of measures of reading performance (Underwood, Briscoe and MacCleary, 1992). These measures were taken from each sentence as it was presented on the screen, and included fixation duration and location, and total reading time for the whole sentence. A number of the sentences, and the reading patterns of a typical reader identified as subject JS, are presented in Figure 2.

This Figure shows the patterns from one of the subjects in the experiment, with the columns representing fixations, and the horizontal lines tracing the paths of saccadic movements. The summary statistics from the experiment are shown in Table 1, which also shows something of the performance of another subject, identified here as HA. This subject, whose data do not contribute to the summary statistics, informed us after the experiment that she had completed a commercial speed reading course. Samples of her reading patterns are presented in Figure 3.

One of the first points from the comparison of the typical readers and HA, using the summary statistics in table 1, is that HA switched her strategy after the first block of sentences. During the second block, she was reading in approximately one quarter of the overall time due partly to making less than half the original number of fixations, and partly to those fixations lasting approximately one third of the duration of those in the first block. The materials were presented in two blocks, with 35 sentences in each. Each of the examples is taken from the second block of sentences when she has adopted a very fast reading style. From HA's reading patterns it is not altogether clear what she was looking at when "reading" these sentences, but it is important to note that her understanding of the passages was superior to that of most of the other readers. HA made very few errors when answering questions about the sentences, suggesting that she was extracting information from the screen rather than relying upon guesswork and fabrication.

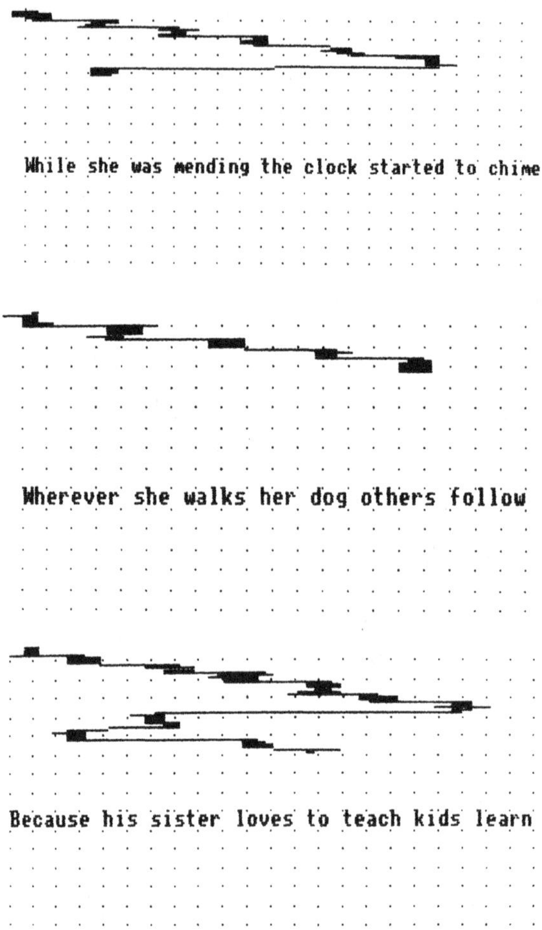

Fig. 2a, b, c. Eye movement patterns of a normal reader
Three sentences and the pattern of eye movements recorded from subject JS, a normal reader of above-average ability, while each sentence was available for inspection. Each pattern begins with a fixation on or about the first letter in the sentence, and each fixation is indicated by a column (time is represented down the page in these print-outs). The horizontal lines between each fixation are saccades, and typically last for 10-30 msec. The average fixation durations for the normal readers are shown in Table 1.

The patterns produced by JS are typical of the majority of readers sampled in the Underwood, Briscoe and MacCleary (1992) experiment. There were successions of mainly left-to-right fixations of slightly less than 300 msec duration, each separated by about 8 character spaces. The right-to-left sweeps towards the end of inspecting the sentences in Fig 3a and 3c are regressive movements produced by a requirement to re-read part of the sentence, possibly to check the original understanding or possibly to re-compute the understanding once all of the words have been recognised.

Table 1. Reading sentences: typical readers and a speed reader.
Summary statistics from the experiment reported by Underwood, Briscoe and MacCleary (1992), in which adults read sentences. The data marked 'Typical Readers' were collected from 30 undergraduates. The sentences either required early closure (eg "While Pam was washing the baby started to cry") or late closure (eg "Though Ben read the book the baby bothered him"), and either had no punctuation or had a comma inserted to indicate the point of clause closure (eg "While Pam was washing, the baby started to cry"). The filler sentences were included in the presentations to facilitate text cohesion and to separate critical sentences. The data from the speed reader (subject HA) have been separated into the two blocks of trials to indicate the strategy switch between blocks.
Measures taken were: sentence reading time (mean time between display of a sentence and the reader indicating comprehension by a button press); number of fixations per line (excluding the initial fixation on the cursor shown immediately before the sentence); average fixation duration (excluding the initial fixation and the terminal wrap-up fixation); and wrap-up fixation duration.

	Early Closure Unpunctuated	Early Closure Punctuated	Late Closure Unpunctuated	Late Closure Punctuated	Filler Sentences
Sentence Reading Time (msec)					
Typical Readers	3126	2600	2941	2684	2560
HA: Block 1	1482	1410	1446	1532	1288
HA: Block 2	404	360	340	388	386
Number of Fixations/Sentence					
Typical Readers	9.7	7.8	9.2	8.2	6.2
HA: Block 1	3.0	2.4	2.4	2.8	2.5
HA: Block 2	1.4	1.0	0.4	0.4	1.4
Average Fixation Duration (msec)					
Typical Readers	283	281	276	277	285
HA: Block 1	239	274	228	261	225
HA: Block 2	79	80	85	90	96
Average Wrap-Up Fixation Duration (msec)					
Typical Readers	272	291	270	270	291
HA: Block 1	318	318	300	298	277
HA: Block 2	198	190	204	256	190

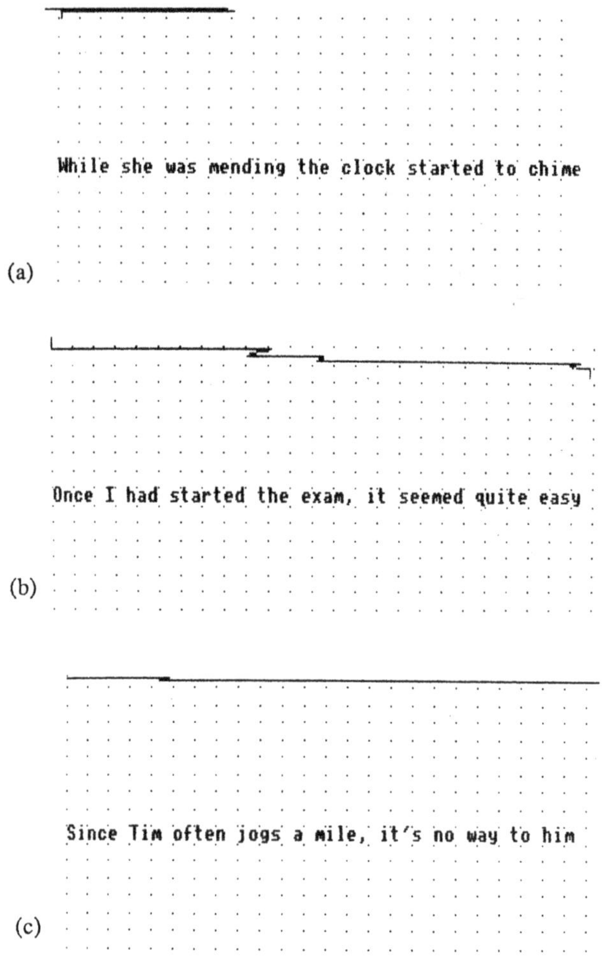

Fig. 3a, b, c. Eye movement patterns of a speed reader
Three inspection patterns produced by subject HA. During the first block of sentences her patterns were similar to those shown in Fig 2, and those in Fig 3 are taken from the second block, when her reading accelerated. There was no typical inspection pattern during this block of sentences. Sometimes there would be a very brief fixation at the beginning of the sentence (Fig 3a), followed by a long saccade to the centre of the sentence, a second brief fixation, and then a long regressive saccade back to the beginning, and a button-press to indicate that the sentence had been read. In Fig 3b the pattern is more similar to the patterns in Fig 2, but with longer saccades, fewer fixations, and briefer fixations, and in Fig 3c a brief initial fixation is followed by a relatively short saccade and a brief fixation, followed by a very long saccade to the end of the sentence, and then the button-press. The summary statistics for HA are presented in Table 1, but as her performance in Block 2 was so variable the means should be interpreted with caution.

Although the most dramatic patterns come from the second block of trials, the data in Table 1 suggest that she was already reading atypically during the first block. Her sentence reading times were below 50% of those of the typical reader, achieved mainly through a reduced number of fixations per sentence. The durations of those fixations were only slightly lower than those of typical readers at this stage. After the break the number of fixations was reduced even further, and the durations were reduced enormously.

How has HA achieved her rapid reading performance? Perhaps she is able to make greater use of peripheral and parafoveal vision than normal readers, or perhaps she has an unusually long persistence of vision which allows for the extraction of information from representations of the sentences after they have been removed from the screen. An alternative explanation takes into account a reported trade-off between reading speed and comprehension. A glance at the literature on speed reading suggests that HA's performance must be at the expense of comprehension, but she made very few errors in answering the questions about the sentences. In their study of speed readers, skimmers, and normal readers, Masson (1974) and Just and Carpenter (1987) confirmed the enormous variation in reading rate suggested by HA's atypical performance, with speed readers fixating about 33% of all words (normal readers: 64%), reading at a rate of about 700 words per minute (normal readers: 240 words per minute), and having gaze durations of about 233 msec (normal readers: 330 msec). Their speed readers were performing as HA did in the first block of sentences which she read, but were quite slow in comparison with HA's performance with the second block. Masson (1974) and Just and Carpenter (1987) also found that speed readers were disadvantaged in answering both high-level questions about the gist of a story, and low-level questions about specific details. On one particular passage taken from Reader's Digest, speed readers answered about 65% of the high-level questions correctly, in comparison to about 80% for the normal readers. With the questions about low-level detail, the speed readers answered about 30% correctly, and the normal readers about 50%. Similar ratios were observed with a passage from Scientific American. Our speed reader, in contrast, out-performed most of the normal readers. One hypothesis, then, is that our questions probed comprehension insufficiently. The questions were not designed to do anything more than ensure that the passages were read for meaning, and it is entirely possible that HA was able to guess the correct answers on the basis of having extracted a few content words from each sentence. Given the nature of the syntax of some of our sentences this does not seem very plausible, but we have no grounds for dismissing the hypothesis and there is good reason from other studies for us to believe that, in general, speed reading is achieved at the expense of comprehension.

However HA has achieved her rapid reading rate, her performance provides a special challenge to the eye-mind assumption. If we observe her eyes while she inspects sentences, then we will learn little about the cognitive processes necessary for comprehension. The direction of her eyes tells us little about the contents of her mind. Only by arguing that HA is not "reading" the sentences can the assumption survive.

22. Conclusions

The strict forms of the eye-attention assumption and the eye-mind assumption do not stand up well to the evidence. Attention is not locked onto the direction of gaze continuously, and observers can process events away from fixation. These parafoveal events may even be used by the eye guidance mechanism in the selection of the next fixation location.

The eye-attention assumption holds that, by recording the movements of the eye, an external observer may have access to the current contents of conscious processing. The assumption requires that we are unable to attend other than where we are fixating, and this can be shown to be false. With stationary eyes, observers can attend a few degrees away from fixation, in order to make use of advance information about the location of target events. With saccadic eye-movements permitted, attention can be shown to move ahead of the eyes. These demonstrations of eye-movements and independent attention-movements across a two-dimensional space are further qualified when the dimension of depth is added to the display. Under very specific conditions attention can be switched from one scene to another without any ocular indication of a movement. Although this is a special case which does not have a general influence upon any working acceptance of the eye-attention assumption, it provides further evidence for doubt.

For the eye-mind assumption to hold, the evidence would need to demonstrate that all processing of an object is completed during its inspection. Carpenter and Just (1983) allow for the continued processing of previously encountered words, but words ahead of fixation also influence the current fixation behaviour. The assumption requires that there are no influences of non-fixated words during reading, and this can be shown to be false. The most useful evidence concerns influences upon eye guidance. Readers do not fixate every word when comprehending text, and the words which are skipped tend to be uninformative in some way. They may be highly predictable from the context of the passage, or they may be from predictable syntactic classes. Words which are not fixated can, nevertheless, influence performance, both in natural reading tasks and in tachistoscopic recognition tasks. Furthermore, readers' eyes move towards words which have been primed by associates, and towards parts of words which are most informative.

The eye-mind assumption accounts for a large proportion of the variance in fixation durations, but it is incomplete, and it does not account for the evidence of parafoveal processing which suggests that we know what to avoid and that we know where to look next.

References

Allport, D.A. (1979). Conscious and unconscious cognition: a computational metaphor for the mechanism of attention and integration. In: *Perspectives in Memory Research*, (Ed) L.G. Nilsson. Hillsdale: Erlbaum.

Balota, D.A., Pollatsek, A. and Rayner, K. (1985). The interaction of contextual constraints and parafoveal visual information in reading. *Cognitive Psychology, 17*, 364-390.

Balota, D.A. & Rayner, L. (1983). Parafoveal visual information and semantic contextual constraints. *Journal of Experimental Psychology: Human Perception and Performance, 9*, 726-38.

Blanchard, H.E., McConkie, G.W., Zola, D. and Wolverton, G.S. (1984). Time course of visual information utilization during fixations in reading. *Journal of Experimental Psychology: Human Perception and Performance, 10*, 75-89.

Blanchard, H.E., Pollatsek, A. and Rayner, K. (1989). The acquisition of parafoveal word information in reading. *Perception and Psychophysics, 46*, 85-94.

Bouma, H. & de Voogd, A.H. (1974). On the control of saccades in reading. *Vision Research, 14*, 273-84.

Broadbent, D.E. (1971). *Decision and Stress*. London: Academic Press.

Broadbent, D. E. (1982). Task combination and selective intake of information. *Acta Psychologia, 50*, 253-90.

Carpenter, P.A. and Daneman, M. (1981). Lexical retrieval and error recovery in reading: A model based on eye fixations. *Journal of Verbal Learning and Verbal Behavior, 20*, 137-160.

Carpenter, P.A. & Just, M.A. (1983). What your eyes do while your mind is reading. In: *Eye Movements in Reading: Perceptual and Language Processes*, ed. K. Rayner. New York: Academic Press.

Carroll, P. and Slowiaczek, M.L. (1986). Constraints on semantic priming in reading: A fixation time analysis. *Memory and Cognition, 14*, 509-522.

Coeffe, C. and O'Regan, J.K. (1987). Reducing the influence of nontarget stimuli on saccade accuracy: Predictability and latency effects. *Vision Research, 27*, 227-240.

Crowder, R.G. (1982). *The Psychology of Reading: An Introduction*. New York: Oxford University Press.

Dixon, N.F. (1981). *Preconscious Processing*. Chichester: Wiley.

Ehrlich, S.F. & Rayner, K. (1981). Contextual effects on word perception and eye movements during reading. *Journal of Verbal Learning and Verbal Behavior, 20*, 641-55.

Duffy, S.A. and Rayner, K. (1990). Eye movements and anaphor resolution: Effects of antecedent typicality and distance. *Language and Speech, 33*, 103-119.

Duffy, S.A., Morris, R.K. and Rayner, K. (1988). Lexical ambiguity and fixation times in reading. *Journal of Memory and Language, 27*, 429-446.

Ehrlich, K. and Rayner, K. (1983). Pronoun assignment and semantic integration during reading: Eye movements and immediacy of processing. *Journal of Verbal Learning and Verbal Behavior, 22*, 75-87.

Everatt, J. and Underwood, G. (1992). Parafoveal guidance and priming effects during reading: a special case of the mind being ahead of the eyes. Manuscript submitted.

Ferreira, F. and Clifton, C. (1986). The independence of syntactic processing. *Journal of Memory and Language, 25*, 348-368.

Ferreira, F. and Henderson, J.M. (1990). Use of verb information in syntactic parsing: Evidence from eye movements and word-by-word self-paced reading. *Journal of Experimental Psychology: Learning, Memory and Cognition, 16*, 555-568.

Fisher, D.F. (1975). Reading and visual search. *Memory and Cognition, 3*, 197-209.

Fisher, D.F. & Shebilske, W.L. (1985). There is more than meets the eye than the eye-mind assumption. In: *Eye Movements and Human Information Processing*, ed. R. Groner, G. McConkie & C. Menz. Amsterdam: North-Holland.

Frazier, L. and Rayner, K. (1982). Making and correcting errors during sentence comprehension: Eye movements in the analysis of structurally ambiguous sentences. *Cognitive Psychology, 14*, 178-210.

Gibson, E.J. and Levin, H. (1975). *The Psychology of Reading*. Cambridge, Mass: MIT Press.

Grindley, G.C. & Townsend, V. (1968). Voluntary attention in peripheral vision, and its effects on acuity and differential thresholds. *Quarterly, Journal of Experimental Psychology, 20*, 11-9.

Helmholtz, H. von (1866). *Helmholtz's Physiological Optics*. Translated from the third edition (1909-1911) by J.P. Southwell (ed.), Rochester, New York: Optical Society of America.

Hochberg, J. (1970). Components of literacy: speculations and exploratory research. In: *Basic Studies in Reading*, ed. H Levin & J.P. Williams. New York: Basic Books.

Hogaboan, T.W. (1983). Reading patterns in eye movement data. in: *Eye Movements in Reading: Perceptual and Language Processes*, ed. K. Rayner. New York: Academic Press.

Hyona, J., Niemi, P., and Underwood, G. (1989). Reading long words embedded in sentences: informativeness of word parts affects eye movements. *Journal of Experimental Psychology: Human Perception and Performance, 15*, 142-152.

Inhoff, A.W. (1987). Parafoveal word perception during eye fixations in reading: Effects of visual salience and word structure. In: *Attention and Performance XII: The Psychology of Reading*, ed. M. Coltheart, London: LEA.

Inhoff, A.W. and Rayner, K. (1986). Parafoveal word processing during eye fixations in reading: Effects of word frequency. *Perception and Psychophysics, 40*, 431-439.

James, W. (1890). *Principles of Psychology*. New York: Holt.

Jennings, G.D. and Underwood, G. (1984). The influence of parafoveal information in a simple reading test. In: *Theoretical and Applied Aspects of Eye Movement research*, ed. A.G. Gale & F Johnson. Amsterdam: North-Holland.

Just, M.A. and Carpenter, P.A. (1978). Inference processes during reading: Reflections from eye fixations. In: *Eye Movements and the Higher Psychological Functions*, ed. J.W. Senders, D.F. Fisher, and R.A. Monty. Hillsdale, N.J.: LEA.

Just, M.A. & Carpenter, P.A. (1980). A theory of reading: from eye fixations to comprehension. *Psychological Review, 87*, 329-54.

Just, M.A. and Carpenter, P.A. (1987). *The Psychology of Reading and Language Comprehension*. Newton, Mass: Allyn and Bacon.

Just, M.A., Carpenter, P.A. and Wooley, J.D. (1982). Paradigms and processes in reading comprehension. *Journal of Experimental Psychology: General, 111*, 228-238.

Kennedy, A. (1978). Reading sentences: some observations on the control of eye movements. In: *Strategies of Information Processing*, ed. G Underwood. London: Academic Press.

Kennedy, A. (1987). Eye movements, reading skill and the spatial code. In: *Cognitive Approaches in Reading*, eds. J. Beech and A. Colley. Chichester: Wiley.

Kerr, J.S. and Underwood, G. (1984). Fixation time on anaphoric pronoun decreases with congruity of reference. In: *Theoretical and Applied Aspects of Eye Movement Research*, eds. A.G. Gale and F. Johnson. Amsterdam: North-Holland.

Kerr, J.S. and Underwood, G. (1985). Comparable priming effects obtained in lexical decision, naming and reading tasks. In: *Eye Movements and Human Information Processing*, ed. R. Groner, G. McConkie, & C. Menz. Amsterdam: North-Holland.

Kliegl, R., Olson, R.K. and Davidson, B.J. (1983). On problems of unconfounding perceptual and language processes. In: *Eye Movements in Reading: Perceptual and Language Processes*, ed. K Rayner. New York: Academic Press.

Kolers, P.A. (1969). Voluntary attention switching between foresight and hindsight. *Quarterly Progress Report of the Research Laboratory of Electronics*, MIT, *92*, 381-5.

Lima, S.D. (1987). Morphological analysis in sentence reading. *Journal of Memory and Language, 26*, 84-99.

Lima, S.D. and Inhoff, A.W. (1985). Lexical access during eye fixations in reading: Effects of word-initial letter sequences. *Journal of Experimental Psychology: Human Perception and Performance, 11*, 272-285.

McClelland, J.L. (1987). The case for interactionism in language processing. In: *Attention and Performance XII: The Psychology of Reading*, ed. M. Coltheart, London: LEA.

McConkie, G.W., Zola, D., Blanchard, H.E. and Wolverton, G.S. (1982). Perceiving words during reading: Lack of facilitation from prior peripheral exposure. *Perception and Psychophysics, 32*, 271-281.

Marcel, T. (1983). Conscious and unconscious perception: experiments on visual masking and word recognition. *Cognitive Psychology, 15*, 197-237.

Masson, M.E.J. (1984). Rapid reading processes and skills. In: *Reading Research: Advances in Theory and Practice*. G.E. MacKinnon and T.G. Waller. Orlando: Academic Press.

Morton, J. (1969). Interaction of information in word recognition. *Psychological Review, 76*, 165-78.

Murray, W.S. and Kennedy, A. (1988). Spatial coding in the processing of anaphor by good and poor readers: Evidence from eye movement analysis. *Quarterly Journal of Experimental Psychology, 40A*, 693-718.

O'Brian, E.J., Shank, D.M., Myers, J.L. and Rayner, K. (1988). Elaborative inferences during reading: Do they occur on-line? *Journal of Experimental Psychology: Learning, Memory and Cognition, 14*, 410-420.

O'Regan, J.K. (1979). Saccade size in reading: evidence for the linguistic control hypothesis. *Perception and Psychophysics, 25*, 501-9.

O'Regan, J.K. (1984). How the eye scans isolated words. In: *Theoretical and Applied Aspects of Eye Movement Research*, ed. A.G. Gale & F. Johnson. Amsterdam: North-Holland.

Olson, R.K., Kliegl, R. and Davidson, B.J. (1983). Eye movements in reading disability. In: *Eye Movements in Reading: Perceptual and Language Processes*, ed. K. Rayner. New York: Academic Press.

Posner, M.I., Nissen, M.J. and Ogden, W.C. (1978). Attended and unattended processing modes: the role of set for spatial location. In: *Modes of perceiving and processing information*, ed. H.L. Pick & J.J. Saltzman. Hillsdale: Erlbaum.

Posner, M.I. and Snyder, C.R.R. (1975). Attention and cognitive control. In: *Information Processing and Cognition*: The Loyola Symposium, ed. R.L. Solso. Hillsdale: Erlbaum.

Rayner, K. (1975). The perceptual span and peripheral cues in reading. *Cognitive Psychology, 7*, 65-81.

Rayner, K. (1977). Visual attention in reading: eye movements reflect cognitive processes. *Memory and Cognition, 4*, 443-8.

Rayner, K. (1984). Visual selection in reading, picture perception, and visual search: A tutorial review. In: *Attention and Performance X*, ed. H. Bouma & D.G. Bouwhuis. Hillsdale: Erlbaum.

Rayner, K. (1986). Eye movements and the perceptual span in beginning and skilled readers. *Journal of Experimental Child Psychology, 41*, 211-236.

Rayner, K., Carlson, M. and Frazier, L. (1983). The interaction of syntax and semantics during sentence processing: Eye movements in the analysis of semantically biased sentences. *Journal of Verbal Learning and Verbal Behavior, 22*, 358-374.

Rayner, K. and Duffy, S.A. (1986). Lexical complexity and fixation times in reading: Effects of word frequency, verb complexity and lexical ambiguity. *Memory and Cognition, 14*, 191-201.

Rayner, K. and Pollatsek, A. (1981). Eye movement control during reading: Evidence for direct control. *Quarterly Journal of Experimental Psychology, 33A*, 351-373.

Rayner, K. and Pollatsek, A. (1987). Eye movements in reading. A tutorial review. In Coltheart, M. (ed). *Attention and Performance XII: The Psychology of Reading*. London: LEA.

Rayner, K. and Pollatsek, A. (1989). *The Psychology of Reading.* Englewood Cliffs: Prentice-Hall.

Remington, R.W. (1980). Attention and saccadic eye movements. Journal of Experimental Psychology: *Human Perception and Performance, 6,* 726-44.

Salthouse, T.A. and Ellis, C.L. (1980). Determinants of eye fixation duration. *American Journal of Psychology, 93,* 207-234.

Schustack, M.W., Ehrlich, S.F. and Rayner, K. (1987). Local and global sources of contextual facilitation in reading. *Journal of Memory and Language, 26,* 322-340.

Shebilske, W. (1975). Reading eye movements from an information-processing point of view. In: *Understanding Language,* ed. D. Massaro. New York: Academic Press.

Shepherd, M., Findlay, J.M., and Hockey, R.J. (1986). The relationship between eye movements and spatial attention. *Quarterly Journal of Experimental Psychology, 38A,* 475-91.

Slowiaczek, M.L. (1983). What does the mind do while the eyes are gazing? In: *Eye Movements in Reading: Perceptual and Language Processes,* ed. K. Rayner. New York: Academic Press.

Taft, M. and Forster, K.I. (1975). Lexical storage and retrieval of prefixed words. *Journal of Verbal Learning and Verbal Behavior, 14,* 638-647.

Taraban, R. and McClelland, J.L. (1988). Constituent attachment and thematic assignment in sentence processing: Influences of content-biased expectations. *Journal of Memory and Language, 27,* 597-632.

Taylor, E.A. (1957). The spans: perception, apprehension and recognition as related to reading and speed reading. *American Journal of Ophthalmology, 44,* 501.

Tinker, M.A. (1958). Recent studies of eye movements in reading. *Psychological Bulletin, 55,* 215-31.

Underwood, G. (1976). Semantic interference from unattended printed words. *British Journal of Psychology, 67,* 327-38.

Underwood, G. (1981). Lexical recognition of embedded unattended words: Some implications for reading processes. *Acta Psychologica, 47,* 267-83.

Underwood, G. (1982). Attention and awareness in cognitive and motor skills. In: *Aspects of Consciousness, Volume 3,* ed. G Underwood. London: Academic Press.

Underwood, G. (1985). Eye movements during the comprehension of written language. In: *Progress in the Psychology of Language, Volume 2,* ed. A.W. Ellis. London: Erlbaum.

Underwood, G., Bloomfield, R., and Clews, S. (1989). Information influences the pattern of eye fixations during sentence comprehension. *Perception, 17*, 267-278.

Underwood, G., Briscoe, T., and MacCleary, S. (1992). How obligatory are parsing strategies? Manuscript submitted.

Underwood, G., Clews, S. and Everatt, J. (1990). How do readers know where to look next? Local information distributions influence eye fixations. *Quarterly Journal of Experimental Psychology, 42A*, 39-65.

Underwood, G., Clews, S. and Wilkinson, H. (1989). Eye fixations are influenced by the distribution of information within words. *Acta Psychologica, 72*, 263-280.

Underwood, G., Hubbard, A. and Wilkinson, H. (1990). Eye fixations predict reading comprehension: The relationship between reading skill, reading speed and visual inspection. *Language and Speech, 33*, 69-81.

Underwood, G. and Thwaites, S. (1982). Automatic phonological coding of unattended printed words. *Memory and Cognition, 10*, 434-42.

Underwood, N.R. and Zola, D. (1986). The span of letter recognition of good and poor readers. *Reading Research Quarterly, 21*, 6-19.

Vonk, W. (1984). Pronoun comprehension. In: *Theoretical and Applied Aspects of Eye Movement Research*, eds. A.G. Gale and F. Johnson. Amsterdam: North-Holland.

The Role of Eye Movements in Perceptual Processes
E. Chekaluk and K.R. Llewellyn (Editors)

SACCADIC SUPPRESSION: A FUNCTIONAL VIEWPOINT

EUGENE CHEKALUK and KEITH R. LLEWELLYN

1. Introduction

A phenomenon associated with voluntary saccadic eye movements that was first described almost a century ago (Dodge, 1900) and which still merits investigation is commonly labelled "saccadic suppression". As we shall later see, this term is in one sense misleading, and may also refer to a number of related, but distinct phenomena related to saccadic eye movements. The following aspects have been described:

1.1 Saccadic suppression and saccadic omission: effects on visual sensitivity.

The loss of visual sensitivity associated with the making of a saccadic eye movement was described early in the history of psychology, but was first given the label "saccadic suppression" by Zuber, Crider and Stark (1964). The basic question associated with this phenomenon is what is the mechanism that is responsible for this drop in sensitivity. The typical approach of the early experimental researchers (e.g. Latour, 1962; Volkmann, 1962) was to record voluntary saccadic eye movements, and to measure sensitivity by presenting a probe (target) during the saccadic sequence. Such a probe was of course merely representational of some visual stimulus that might impinge on the retina during the saccade. An alternative method of measuring visual sensitivity during saccades was first used by Campbell and Wurtz (1978). These researchers omitted the probe from the saccadic sequence and instead posed the question as to why the retinal blur that was produced when a saccade was made was not perceived. They termed this phenomenon "saccadic omission". It should be noted that if the focus of concern in the saccadic situation is the loss of visual sensitivity associated with it, then both of these terms "saccadic suppression" and "saccadic omission" refer to the same phenomenon. The fact that they are different merely reflects a difference in methodology used to measure sensitivity. Both methodologies have yielded an impressive set of data about the characteristics of visual sensitivity loss associated with saccades as well as providing support for the various theories put forward to account for this loss. These will be dealt with in Part 2. of this chapter.

1.2 Loss of perceptual stability.

Whenever a saccade is made, one scene is substituted on the retina for another within a very short space of time. Given that visual sensitivity is reduced during the period of the eye movement (see above) such a substitution would presumably have an affect on what is known as visual direction constancy, which is the invariance of apparent egocentric direction despite changes in oculocentric direction. Shebilske (1977) demonstrated that it is only the dynamic component of the constancy that is subject to breakdown in these circumstances and not the static component. A number of theoretical approaches have been taken to explain why under normal viewing conditions perceptual stability is maintained. Matin (1986) relates saccadic suppression to perceptual stability by proposing two explanations: retinal filtering by means of metacontrast, and a central component which has as its basis spatial uncertainty. (It should be noted that the whole question of perceptual stability despite eye movements involves phenomena other than saccadic suppression e.g. visual persistence, which do not concern us here).

1.3 Saccadic suppression of displacement.

A phenomenon associated with voluntary saccades that has been described only relatively recently is the saccadic suppression of displacement. This was first named by Stark, Kong, Schwartz, Hendry and Bridgeman (1976) and basically refers to an increase in threshold for detection of a change in stimulus location during saccades. Most recently Li and Matin (1990a,1990b) have extensively studied this phenomenon and concluded that the mechanism responsible for it involves an extraretinal signal which denotes eye position, and that furthermore the retinal stimulus itself does not contribute to this effect.

It should be noted that although we have made a distinction above about the three uses of the term saccadic suppression, they tend to be related and have been integrated in terms of theory. For example, it is clear that those that emphasise the role of an extraretinal signal in the saccadic suppression of visual sensitivity (e.g. Riggs, Merton and Morton, 1974; Matin and Matin, 1972) would also emphasise such a signal in determining visual direction despite eye movements (Matin, 1986) and in being responsible for saccadic suppression of displacement (Li and Matin, 1990a).

In the discussion that follows we will concentrate on saccadic suppression in the first sense mentioned above, namely the phenomenon associated with visual sensitivity. Even though we have argued that both saccadic suppression and saccadic omission refer to the same phenomenon measured in different ways, the structure of our discussion will be to treat the results and theoretical implications of each approach separately before presenting an overall approach to the question, as the two approaches seem to have generated research which in general does not overlap.

2. Saccadic Suppression: Explanations and Theories.

The primary focus of interest of studies using this approach has been to provide evidence for or against some form of extraretinal signal as the source of the drop in visual sensitivity associated with saccades. Although the debate between proponents of both extreme positions has been at times heated (see Riggs, 1976 and Brooks, Impleman and Lum, 1981), the present position seems to be one of agreement that this form of saccadic suppression involves both an extraretinal component and a retinal one (e.g. see Bridgeman, 1977; Chekaluk and Llewellyn, 1990; Li and Matin 1990a). Given that this is the case, the term "saccadic suppression" can be seen to be rather unfortunate, as it seems to exclude the possibility of retinal factors being involved in the phenomenon. (A "suppression" presumably involves an active "suppressor" which logically could only be some centrally located mechanism operating via a corollary discharge). Even though both factors are now seen to contribute to the phenomenon, the emphasis that each receives is certainly driven by the theoretical position adopted. Stimulus contributions have been assessed as "of relatively trivial interest" (Alpern, 1969) and even in the earliest days Dodge felt that if saccadic suppression could indeed be explained simply as a result of retinal events it would "rob the problem of most of its mystery" (Dodge, 1905, p. 199). On the other hand, stimulus contributions have been defended by Matin who stated that "it (i.e. saccadic retinal stimulation) is itself a mysterious object and....Nature may well reward us if we explore it as such and do not regard it primarily as an artifact". (Matin, 1976, p. 115). In the following sections we will examine in some detail both the role of the extraretinal signal and the retinal stimulus in saccadic suppression studies.

2.1 The role of the extraretinal signal

From its earliest descriptions saccadic suppression has been used as a means of investigating possible neural mechanisms involved in visual perception. From a neurophysiological perspective, a search was undertaken for cells in the visual pathway that would change their firing rate only when a voluntary saccade was made. Such cells would then be implicated as the source of the extraretinal signal involved in producing saccadic suppression. A cellular response of this type has come to be known as a corollary discharge. A large number of brain areas were examined for cells which might provide evidence of a unique corollary discharge corresponding to voluntary saccades by Jeannerod, Kennedy and Magnin (1979). The areas studied were the dorsal nucleus of the lateral geniculate body, the visual cortex, the ventral nucleus of the lateral geniculate body, the pregeniculate nucleus and the retinotectal pathway. No evidence of a unique corollary discharge was found by these authors. Richmond and Wurtz (1980), on the other hand found evidence of a corollary discharge mechanism in the superior colliculus of the monkey and Zaretsky (1982) of a similar mechanism in the descending contralateral movement detector of the locust. The evidence of the neurophysiological studies mentioned was, however, ambiguous. In all cases, the corollary discharge was also activated in the absence of voluntary saccades, when stimuli in the animal's visual field were moved with saccadic velocities in front of the stationary eye (Wurtz, Richmond and Judge, 1980; Zaretsky, 1982). It would seem, then, that these data suggest that under the viewing conditions used by the above researchers, no *unique* corollary discharge mechanism exists. It is in fact the viewing conditions used, and their control, which has generated the most research in psychophysical studies of the extraretinal signal.

The first indication that the phenomenon of saccadic suppression of visual sensitivity could be used to investigate the presence of an extraretinal signal involved with eye movements came from the plot of the timecourse of the phenomenon. Latour (1962) pioneered the technique of presenting a probe to a saccading eye and measuring detection probability. Although his studies were not controlled in the sense that saccade magnitude and duration were free to vary, subsequent controlled studies (Volkmann, Schick and Riggs, 1968) report the same salient finding: visual sensitivity begins to drop *prior* to the commencement of the saccade. This finding ties in neatly with the idea of the extraretinal signal - the beginning of the drop in visual sensitivity coincides with the centrally driven command to initiate the eye movement. Once the eye movement has commenced visual sensitivity is substantially reduced and the visual input has no chance to disrupt perception of a stable world. The above finding concerning the timecourse of the suppression is of course merely necessary, but not sufficient, to support the existence of an extraretinal signal.

A necessary and sufficient condition for the existence of an extraretinal signal would be a demonstration that saccadic suppression of visual sensitivity occurs in the absence of retinal stimulation. Attempts to make such a demonstration were undertaken early in the history of experimental research on saccadic suppression. Both Volkmann (1962) and Volkmann, Schick and Riggs (1968) controlled their experimental situation in such a way as to eliminate all contour from the visual field, as well as limiting any smear that might be produced on the retina by a probe stimulus. Under these conditions, there is a consistent finding of an increase in visual threshold of .5 log units associated with saccades. It may be argued that because of technical limitations, these early studies did not fully eliminate retinal stimulus input. Such an argument does not apply to an experimental innovation used by Riggs, Merton and Morton (1974) and subsequently extended by Riggs (1976). This technique involved using conditions of total darkness where the experimental subject would make voluntary saccades and would be presented with visual phosphenes. Under these conditions the researchers report a small, but significant increase in visual thereshold. The possibility that the sensations produced by phosphenes (which are diffuse) might be quantitatively different to those produced by conventional visual stimuli was raised by Matin (1974). As a response to this Riggs and his co-workers eliminated all contour by using a Ganzfeld as well as placing diffusers over the eyes. Any stimulus input, even that which in other studies might have been produced by fixation guides or the nose and cheeks of the observer was thus eliminated. The results obtained were similar to those obtained using phosphenes.

Before examining the role of retinal factors involved in saccadic suppression in the next part of the chapter, the magnitude of the extraretinal effect reported requires some elaboration. A survey of the saccadic suppression literature reveals a wide variety of conditions used, and thus not surprisingly, a wide variety of reported effect sizes. For example, Uttal and Smith (1968) found a rise in threshold of .2 log units whereas Zuber and Stark (1966) report threshold rises over 1 log unit. In a suppression study which purports to provide evidence of an extraretinal source, it is essential that all steps be taken to remove contour (see discussion of Riggs et al's work above). Under these conditions, there is a concensus that the order of magnitude of the rise in threshold associated with saccades is around .5 log units.

2.2 The Role of Visual Masking

Despite a fair amount of evidence that an extraretinal signal is required for the saccadic suppression of visual sensitivity, there are those researchers that believe that the effect can be totally explained in terms of the retinal stimulus (see for example Brooks and Fuchs,1975; Brooks, Impleman and Lum, 1981). The thrust of the position that the phenomenon can be ascribed to retinal factors derives from two sources. Firstly, there is the theoretical position first formulated by Woodworth (1906) that, provided stimulus conditions were held constant, there should be no difference in perception between the saccading and the stationary eye. An empirical review of this position was recently completed by Sperling (1990). His conclusions were that for small saccades as well as pursuit movements, it is the content of the input which is processed, and not its origin (sensory or motor) at all, but the highest levels, of processing. In other words, Sperling claims that where differences are found on psychophysical tasks that compare voluntary eye movements with imposed image motion, these are primarily due to failing to provide equivalent retinal stimuli under both conditions.

A second source of the position that saccadic suppression is essentially a retinal phenomenon was the work performed by Wurtz (1969a,1969b) using animal subjects and MacKay (1970a,1970b) using humans. Wurtz compared the neuronal responses measured from the visual cortex of monkeys when a saccade was made across an effectively stationary stimulus with those when the same stimulus was moved with saccadic velocity in front of the immobile eye. He found no significant differences in visual sensitivity between these two situations. A similar technique was used by MacKay, who had subjects fixate a screen which was rapidly displaced by means of a mirror system. A flash was presented at various temporal and spatial locations before, during and after this period of displacement, and subjects were required to report the visibility or otherwise of the flash. An interesting characteristic of both these studies (and other subsequent studies using such "saccade analogues") was that the timecourse of the suppression of visual sensitivity obtained was similar to that observed when true saccades were made, namely visual sensitivity was observed to fall *before* the commencement of the displacement of the visual field, and any neuronal measures were similarly affected. These findings seemed to suggest that retinal factors must certainly play a role in the saccadic suppression of visibility. The question remained, however, as to whether such retinal factors could fully explain the suppression effect and secondly by what mechanism these retinal factors operate. As mentioned in the previous section, it appears that most researchers in the area concede that *both* retinal and extraretinal factors operate to suppress visual sensitivity. There are, however, some researchers that believe that retinal factors are sufficient to explain the suppression effect. Brooks and Fuchs (1975), for example, found no evidence of loss of visual sensitivity when contour was absent from the visual field, and interpret this result as ruling out the influence of extraretinal factors. In any case, one result is quite unambiguous - under ordinary viewing conditions retinal factors are responsible for most, if not all observed suppression of visual sensitivity. This is evident since the extraretinal effect seems to be quite small (about .5 log units increase in threshold when potential retinally based influences are eliminated) whereas the suppression effect under ordinary viewing is complete i.e visual input is totally eliminated.

The answer to the second question as to the retinal mechanism implicated in saccadic suppression has generated much research activity. Early attempts to identify the mechanism as "retinal smear" (Dodge, 1905) or "retinal shearing forces" (Richards 1968, 1969) were either not supported by subsequently available data, or proved to difficult to test experimentally. Most contemporary researchers agree that the retinal factor involved in saccadic suppression is visual masking (Breitmeyer, 1984). Visual masking, of course, refers to a broad phenomenon which is sub-classifiable into a wide variety of distinct instances e.g. masking by light (forward or backward), metacontrast etc. (see Kahneman, 1968 for review). The particular form of visual masking most often mentioned in the context of saccadic suppression is metacontrast. Althought strictly speaking this refers to backward masking by adjacent contours (Lefton, 1972), it is clear that most reports mean it to include both backward and forward masking effects. The role of metacontrast as a retinal source of saccadic suppression is well established. In a well-designed experiment Matin, Clymer and Matin (1972) showed that a slit of light illuminated during a saccade would either be seen as smeared, partially smeared or unsmeared depending on its duration and luminance. It is clear from the manipulations of this study that successive retinal stimulation results in an interaction of stimuli such that previously presented stimuli are not perceived. The authors, however, caution against inferring that this is the sole masking possibility in the saccade situation. Such a caution was elaborated by Matin (1974) who clearly states that spatial interactions involving simultaneous presentations of stimuli are also present in saccades and these interactions could also give rise to powerful masking effects. The emphasis on metacontrast in saccadic suppression has led to the relative neglect of such "spatial" interactions i.e. those interactions that occur as a result of movement of contours across the retina, rather than "temporal" interactions which arise from the fact that the visual image is displaced on the retina as a result of the saccade. It is clear from Matin's statement, however, that such spatial interactions are deserving of attention. MacKay's (1970a, 1970b) work using a fixating eye and moving a field rapidly in front of it raised the possibility that lateral masking effects of moving contours were responsible for the loss of visual sensitivity observed in his experiments. (MacKay did not actually label the effect "masking" but does seem to imply that the movement of the saccadic image contributes to its own loss of visibility.) Subsequently, however, MacKay (1973) makes it clear that he believes that it is the sudden replacement of the structured presaccade field by a structured postsaccade field which is responsible for the effects observed in his earlier experiments. In other words a temporal interpretation is given for these findings.

A large body of work was produced in the 1970s on the possible sources of visual masking by Mitrani and his co-workers. The results that they report are essentially that temporal interactions produced by displacement of the visual field do not produce the masking effect involved in saccadic suppression. For example Mitrani, Radil-Weis, Yakimoff, Mateeff and Bozkov (1975) argue that their results show that it is the movement of contours and not the disappearance of an image from one location and its reappearance in another which is responsible for masking effects in saccadic eye movements. Subsequently, Mateeff, Yakimoff and Mitrani (1976) conclude that it is not clear from their research what feature of the masking stimulus during the movement period contributes to the suppression effect : the continuous motion of the structure across the retina, or the successive change of luminance on different receptors.

In the light of the prior emphasis on temporal effects arising from the pre- and post-saccadic fixations these results are somewhat surprising. One possible explanation for the Mitrani group's results was that they could not assess the effects of the intrasaccadic stimulus independently of the pre- and/or postsaccadic stimulus. Masking effects arising from fixation points, for example, were always present in their studies. In one study where a saccade analogue similar to that used by MacKay was employed, Mateeff et al (1976) did in fact have a necessary factorial structure to assess the masking effects of the moving contour phase independently of the pre and post saccadic structured fields, but such an analysis was not done. The first attempt to isolate the contribution of moving contours to the masking effect in saccades independently of other masking effects was carried out by Llewellyn and his co-workers (McMullen, 1978; Chekaluk (1980); Chekaluk and Llewellyn (1990).

3. Empirical work on the role of masking in saccadic suppression

3.1 The Work of Llewellyn and McMullen

As mentioned earlier, most contemporary researchers agree that it is not the case of either an extraretinal or a retinal basis for saccadic suppression : both systems operate with retinal events seeming to be most heavily implicated under ordinary viewing conditions where contours abound and the potential for masking type interactions is optimal. Given that this is the case, it would appear that the ideal way to assess the retinal contribution to saccadic suppression would be to use a saccade analogue of the type pioneered by MacKay (1970a) and refined by Sperling (1990), as this would of necessity eliminate extraretinal factors. This technique has one further advantage for investigations of the source of visual masking that contributes to the loss of visual sensitivity - it is possible to assess the role of the intrasaccadic or movement phase of the saccadic sequence independently of the effects of the fixations prior to and subsequent to the saccade. In real eye movement studies, of course, the movement phase cannot be manipulated in this way. As mentioned previously, apart from frequent demonstrations that metacontrast plays a role in saccadic suppression, there have been no attempts to evaluate the role, if any, self-masking by moving contours might play in this context. This then was the aim of the first set of investigations by Llewellyn and his team - to test the notion that movement of contours across the retina was the primary determinant of the saccadic suppression effect in the absence of an extraretinal signal.

The basic piece of equipment used in the following experiments was a 3-channel tachistoscope. In the earlier experiments reported this was connected to a timer and tachistoscope controller: the function of these was replaced in the later experiments by computer. The tachistoscope was modified so that one channel, rather than displaying the conventional 6° by 9° stimulus card when illuminated would display a belt which was attached to a motor. The motor rotated the belt at 400° of visual angle per second. If this field of the tachistoscope was illuminated it would expose a section of the belt for the duration of illumination. The following fields were displayed:

1. A homogenous white tachistoscope card with a $.25^{o}$ black fixation spot at its centre.

2. The part of the belt attached to the motor(described above) illuminated at the time. The belt was painted with 1° square gratings oriented vertically.

3. A stationary version of 2. above. 1° gratings oriented vertically were drawn on a standard tachistoscope card. Two such displays were constructed - one where the end grating was white (and thus visible when the appropriate tachistoscope channel was illuminated) which was

in toto 9^{0} wide, and one where the end grating was black (and thus not visible when illuminated) leaving a display field of 7^{0} effective width.

4. Horizontally oriented $.9^{0}$ gratings painted onto a standard tachistosope card. Since both top and bottom gratings were white the effective height of this field was 6^{0}. This was the test stimulus.

Since four stimuli were constructed and only three channels were available in the tachistoscope, the experiment was run under two repeat conditions. In repeat condition one, the gratings in motion (described in 2 above) were used and in repeat condition two stationary gratings were used (described in 3 above). Conditions were randomised across subjects. Since the motor used to drive the grating display was noisy subjects were provided with a set of headphones through which they heard white noise, the level of which was adjusted till they could no longer hear the motor.

A trial consisted of the following sequence of display fields. Firstly the blank field (1) was presented for 1.5s. This was then replaced by the vertical gratings (either moving(2) or stationary(3)) for 50ms. The final display was the blank field (1) for 4s. The trial thus lasted for approximately 5.5s, with a period of darkness of .5s duration between one trial and the next. This served to warn the subject that a new trial was about to begin. For each of the repeat conditions (stationary or moving vertical gratings) the test stimulus (4) was presented at one of three times during the experimental trial:

Time 1: presented for 4ms 31ms prior to the commencement of the vertical gratings display(2 or 3)
Time 2: presented for 4ms immediately prior to the commencement of the vertical gratings display(2 or 3)
Time 3: presented for 4ms 24ms after the commencement of the vertical gratings display(2 or 3)

Under all conditions one display replaced, rather than superimposed the preceding display.

Prior to the commencement of the experiment proper, the acuity of the test stimulus(4) was calibrated for each subject. A 4ms presentation of the test was observed by each subject and its illumination decreased until it was not visible. This enabled a rough calibration of threshold for each subject. Secondly, a small number of experimental trials were presented to the subject where the test stimulus appeared at Time 3 during the experimental sequence, along with a number of catch trials where the test stimulus was not presented. Starting at the rough threshold level for the subject, the intensity of the test stimulus was varied until the subject was responding to it in a forced choice paradigm with a false alarm rate of 20 to 30%. Once this occurred, the intensity level of the test was fixed at this point for the experimental trials proper.

As mentioned earlier, on one half of the trials the vertical gratings were displayed as moving(2) and on the other half of the trials they were displayed as stationary(3). Further, the test stimulus (4) was present on only half the total number of trials in each experiment. For each of the three times (Time 1 to Time 3) therefore, the following trial types occurred with equal frequency:

(1) Vertical gratings moving (2) / test stimulus(4) present [MP]
(2) Vertical gratings moving(2) /test stimulus(4) absent [MA]
(3) Vertical gratings stationary(3) / test stimulus(4) present [SP]
(4) Vertical gratings stationary(3) / test stimulus(4) absent [SA]

Subjects were given 30-40 practice trials, followed by 200 experimental trials where responses were recorded. The 100 vertical gratings moving(2) trials were run as a separate block from the 100 vertical gratings stationary(3) trails, but block order was randomised across subjects.The presence or otherwise of the test stimulus(4) was similarly randomised across trials. The task of the subject was to view the stimulus sequence and respond using a forced choice procedure about the presence of the test stimulus(4). The four categories of response allowed were: "Yes, certain; yes, uncertain; no uncertain; no, certain". Such a method of response collection allows the calculation of standard scores which represent detection values using the Torgerson method of successive intervals (see McNichol,1978) which, being criterion free are regarded as being most appropriate to measuring visual sensitivity in this context (Pearce and Porter, 1970).

Three groups of 30 subjects were used for these experiments. Each group of 30 subjects served as observers for one of the three Time conditions.

Detection values, expressed in standard scores, were obtained for each Time and mean values are presented in Table 1. A repeated measures analysis of variance was carried out to assess both whether subjects were capable of discriminating the test stimulus (i.e. SP-SA, MP-MA) and whether there was a difference in sensitivity between vertical gratings moving(3) and vertical gratings stationary(4) [i.e. (MP-MA)-(SP-SA)]. The appropriate F ratios are presented in Table 2. The limitations of the equipment available at this time did not allow the three Time conditions to be tested appropriately within the one subject. It should be noted, however, that one focus of interest would be to test for differences between these Time conditions. Using the current

Table 1. Mean Detection Values, in Standard Scores, for each Time Condition in the Llewellyn & McMullen Experiment

	MP	MA	SP	SA
Time 1	1.248	-0.218	-0.125	-0.937
Time 2	1.547	-0.181	-0.275	-1.091
Time 3	1.900	-0.479	-0.354	-1.093

Table 2. F Ratios Obtained for all Comparisons in the Llewellyn & McMullen Experiment In all Cases, Critical F = $F_{.05/3;1,29} = 6.46$

	Condition		
	SP-SA	MP-MA	(SP-SA)-(MP-MA)
Time 1	58.53	52.43	9.36
Time 2	62.05	80.64	20.53
Time 3	40.88	119.63	32.58

experimental design such comparisons could not be legitimately made, since they would assume a constant criterion across the Time factor, and as different subjects were used for each Time level, such an assumption was not justifiable.

The results of this experiment were quite clear. An intrasaccadic stimulus which is in motion does not result in a lowering of visual sensitivity. This is in direct contradiction to the findings of Mitrani et al (1975) who claim that the movement of contours across the retina produces masking effects in saccadic eye movements which result in a lowering of visual sensitivity during this period. The results also are in direct contadiction to the interpretation offered by MacKay(1970a) of the effect he observed using a saccade analogue procedure, namely that moving contours that produce the displacement of an image on the retina produce the visual sensitivity loss observed. In fact, when compared to a visual field identical in all respects to the moving one, a stationary field was seen to produce detection values far lower for the test stimulus. Both of these results require some explanation: fortunately only one mechanism is sufficient to explain both. To start with the second result first, namely the reason that the stationary intrasaccadic stimulus resulted in lowered visual sensitivity. When the stationary "intrasaccadic" stimulus was used in the presented sequence, the following displays were presented: Blank;Vertical Gratings;Blank. Detection of the test stimulus (Horizontal gratings) was then measured. In Times 1 and 2, the possibility was there for the vertical gratings to act as a backward mask on the test stimulus, and in Time 3 both backward and forward masking could operate. Even though it is known that masking is at its most effective when gratings are of the same or similar orientations, masking effects are still evident when gratings are oriented at 90° to one another (Campbell and Kulikowski, 1966). The result obtained with the moving gratings can similarly be seen to be due to the effects of visual masking. Again, let us consider the display sequence: Blank; Vertical gratings in motion; Blank. When the vertical gratings are viewed continuously, their appearance is of a field of gray (labelled "gray-out" by Campbell and Wurtz(1978)). Again the test stimulus is a field of horizontal gratings. Even though masking effects from homogenous fields are possible (Stoper and Mansfield, 1978) they are unlikely to be large. Thus the vertical stipes in motion (effectively a "blank") do not act as effective backward (Times 1, 2 and 3) or forward (Time 3) masks, and detection of the test stimulus under the moving gratings condition is significantly better for each Time than that under the stationary gratings stimulus. Both findings, then, simply support the notion that the action of visual masking is indeed relevant in the saccade situation. The question remains, however, as to what exact form of masking is relevant here, and under what conditions optimal masking effects will take place in saccades.

The next step, therefore, in Llewellyn's laboratory was to directly test which form of visual masking was more likely to play a role in saccades. Two possibilities exist. Despite the finding by McMullen that the moving gratings (the intrasaccadic stimulus) were ineffective in masking a structured target, it could be argued that his experiments lacked an appropriate control to extend the argument to say that the relevant aspect in saccadic suppression could not be moving contours. In order to make such a statement, it would be required to compare visual sensitivity under the moving gratings condition used by McMullen with a field that was phenomenologically identical to that produced by the moving gratings but was stationary. It will be recalled that McMullen's control was to simply display vertical gratings that were stationary. It will also be recalled that free inspection of the moving gratings resulted in the perception of a gray field ("gray-out"). The appropriate control, therefore, to enable the evaluation of the possible relevance of the movement aspect in the saccade to contribute to visual sensitivity loss would be to compare a gray-out produced by moving contours to a similar gray-out which was stationary.

3.2 Chekaluk and Llewellyn's Experiments

The first experiment carried out by Chekaluk in Llewellyn's laboratory and reported in Chekaluk and Llewellyn (1990) attempted to directly assess the dynamic aspects of the intrasaccade input on sensitivity. The effects of structure in the extrasaccade fields can concurrently be examined by comparing highly structured extrasaccade fields with those that contain only a fixation point. By combining extra- and intrasaccadic effects factorially the contribution of each can be fully assessed. This would allow a resolution of the conflicting views of the source of masking in the saccade, as well as testing the notion put forward by Matin(1974,1976) that a combination of intra- and extrasaccadic effects would provide the conditions for maximal masking effects.

To a large extent the apparatus and procedure used was similar to that described previously, with the following notable differences. In terms of the stimuli used, the extrasaccadic fields could be as before (i.e. blank with a central fixation dot) or structured, in which case they consisted of .5 cpd horizontal and vertical square wave gratings with a central fixation point. The moving vertical gratings(as before) were used as one of the intrasaccadic displays, but the other intrasaccadic display was a field of stationary uniform midgray. The target stimulus was also different. It was a circular patch of horizontal square wave gratings, 2° in diameter, centred on a tachistoscope card, with the gratings having a frequency of 2.5 cpd. The timing of stimulus presentations was as follows. The pre-saccadic field (either structured or unstructured) was presented for 1.5s, followed by the intrasaccadic field (either moving or stationary) for 50 ms, followed in turn by the post- saccadic field (either structured or unstructured) for 1.5s. On any one trial, the pre- and post- saccadic fields would be identical. Nine subjects were used in the experiment and 1,000 trials (in two blocks of 500) were given. The pre/post saccade field was varied over blocks, and the intrasaccade field and target onset times were varied within blocks. Onset time (relative to intrasaccade field onset) was varied as follows: 80 or 40 ms before onset of the intrasaccade field, or 20, 90 or 130ms after its onset. For the 20 ms after onset condition, the target occupied the middle 10 msec of the 50 msec intrasaccade period. For further details of the methods, subjects and procedure used in this study, the reader is referred to Chekaluk and Llewellyn (1990).

Ratings for each onset time were converted into d'_e values using the Torgerson method as mentioned earlier. Mean d'_e values are graphed in Figure 1. A further statistical analysis revealed that visual sensitivity was significantly lowered when the pre- /post saccade was structured(pattern) compared to unstructured (plain) [$F = 18.6$, $p = .002$]. When the moving intrasaccade field was compared to the stationary version (blank), no significant difference was found [$F = .086$, $p = .777$], nor was there a significant interaction effect between pre-/postsaccade and intrasaccade field conditions [$F = 3.168$, $p = .113$]. The statistical analyses were based on a fully factorial repeated measures design using the MANOVA approach to repeated measures.

Both the statistical analysis carried out and the plotted sensitivity values in Figure 1 show quite clearly that in order to obtain a "saccadic suppression" type curve for visual sensitivity, the vital ingredient is the presence of structure in the pre- and postsaccadic fields, regardless of the presence or absence of rapidly moving contours in the intrasaccade period. Furthermore, there is no evidence of any enhancement effect that moving contours might have on the effects due to the structured fields.

It should further be noted that the stimulus sequence that consisted of structure/blank/structure did not involve any displacement (the blank was not in motion and the two structured fields were identical). This finding

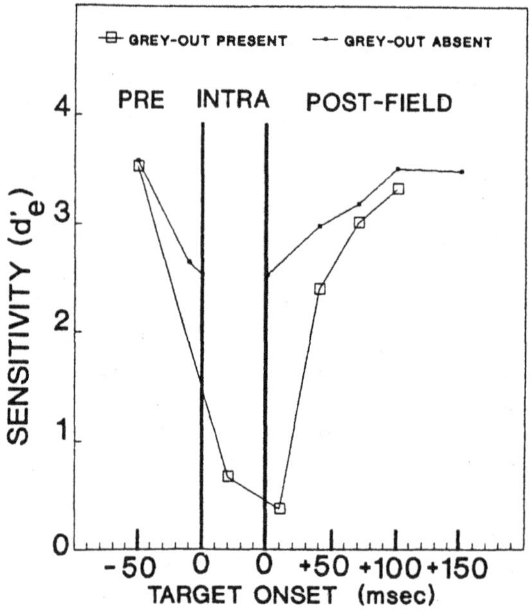

Fig. 1. Sensitivity means for the combinations of structured and nonstructured pre- and postsaccade fixation fields, with intrasaccade fixation fields consisting of either contours moving at a saccadic rate or a stationary gray field. (Figure and quoted caption from Chekaluk & Llewellyn, 1990; reprinted with permission)

rejects the notion that the effect observed was due to either a displacement in the sense of motion of a structured image (MacKay, 1970a) or in the sense of a change in the relative positions of visual field structures (MacKay, 1973).

One possible objection to the results reported in this experiment is that the two times that the target stimulus occurred prior to the intrasaccadic stimulus (80 ms and 40 ms) were sufficiently removed from the onset of that stimulus to question whether we have really shown the charactersitic pre-movement drop in visual sensitivity exhibited in saccadic suppression curves. A follow-up experiment using a smaller number of subjects (4) but with 10 possible occurrence times of the target stimulus during the sequence (including one where the target was presented only 10 ms prior to the onset of the intrasaccadic stimulus) served to confirm the results of the earlier experiment. A simple comparison in this confirmatory experiment that tested the difference in sensitivity to the target stimulus presented 80 ms prior to intrasaccade onset with that presented 10 ms prior proved highly significant ($F = 39.6$, $p = .008$). These results seem to clearly indicate that it is the presence of structured fields, both before and after the intrasaccadic period that result in visual sensitivity loss associated with saccades. The role of the intrasaccadic period has been considered by Chekaluk and Llewellyn (1990) and readers interested in this aspect are directed to that source. For the present, we wish to consider the processes by which the structured pre- and postsaccade fields lead to the typical saccadic suppression sensitivity curve. There is general agreement that the process whereby visual sensitivity drops during saccades made in photopic conditions is a form of visual masking (see Matin, 1974). Both backward (Matin, Clymer and Matin, 1972) and forward (Campbell and Wurtz, 1978) masking have been implicated in the process, although no real systematic attempt has yet been made to directly measure the relative contributions of each type of masking. The next experiment reported addresses these issues.

Both Breitmeyer (1980) and Breitmeyer and Valberg (1979) have argued that the postsaccadic field acts to prevent the persistence of the presaccadic field, thus maintaining the stability of the world during eye movements. One of the implications of this argument is that the backward masking effect of the postsaccadic field must be much stronger than the effect of the presaccadic field in a forward direction, since it is consistently fully masked. As mentioned earlier, Matin, Clymer and Matin (1972) also stress the role of backward masking in this situation. They showed in their experiment that a display lateral to a flash of light shown during a saccade, when shown subsequently was capable of masking the flash. They correctly identify this form of masking as metacontrast (backward masking by non-overlapping contours) and in a subsequent paper Matin (1974) argues that metacontrast evolved in such a way to produce saccadic suppression. She argues, however, that under normal conditions, conditions which would allow for forward as well as backward, overlapping as well as non-overlapping masking exist in the saccadic situation. The emphasis placed upon metacontrast is mainly as a result of its uniqueness, and a need to find its ecological relevance.

In contrast to the abovementioned studies which seem to emphasise the role of backward masking in saccadic suppression, a study carried out by Brooks, Impleman and Lum (1981) showed that the effect of forward masking is greater than that of backward masking in the saccadic situation. Using both real eye movements and then a saccade anlaogue procedure, these researchers altered the luminance of masks which either preceded or followed saccades. Whilst the presaccadic mask was able to affect thresholds to visual targets presented up to 180-200 ms after its offset, a postsaccadic mask had an effect of much shorter duration (100-150 ms). The authors do not offer any explanations for this asymmetry, nor do they attempt to reconcile these findings with the earlier emphasis on backward masking as the prime effect in this situation.

Using the saccadic omission rather than the saccadic suppression approach, Campbell and Wurtz(1978) also examined the difference between forward and backward masking in this situation. Their findings were that if either that pre- or post- saccade field was illuminated for approximately 100 ms, then no blur or gray-out was perceived. There were no differences in the degree of potency of masking between forward and backward masks. In a subsequent paper by this team of researchers (Corfield, Frosdick and Campbell, 1978), the earlier experiments were replicated using saccade analogues and similar results obtained. This finding suggests that forward masking is as effective as backward masking in producing saccadic omission, a point made by Campbell and Wurtz (1978, p 1302). The masking interpretation of these results was subsequently abandoned by Corfield et al (1978) since the stimulus masked was a featureless gray field, and they argued that masking was not the appropriate term to describe its elimination. The finding of Stoper and Mansfield (1978) that blank featureless fields are indeed masked, however, supports the original claim by Campbell and Wurtz that they did demonstrate masking effects where forward masking was equally as strong as backward.

The studies, then, which have attempted to assesss the relative contribution of forward and backward masking in the saccade situation have produced results which are equivocal. One of the possible reasons for this is that in most cases the findings concerning masking were either incidental to the main aim of the study, or the experiemntal situation used was such that only one type of masking was possible. The two experiments reported here attempt to directly measure both the extent and the relative strengths of backward and forward masking in saccade analogues. Both the traditional "saccadic suppression" and Campbell and Wurtz's "saccadic omission" approaches will be taken.

One group of ten volunteer subjects were tested. Presentation of stimuli was by means of the tachistoscope previously described. Three stimuli were used. The test stimulus was a 1^{0} diameter circular patch of gratings, each $.2^{0}$ wide oriented horizontally. Top and bottom gratings were both black. The gray stimulus was the featureless field of homogenous grey described earlier. The masking stimulus was a field of 1^{0} wide gratings oriented vertically. Extreme gratings were both white. The sequence of stimulus presentattion was as follows. The gray was shown for 20 ms, followed by the test stimulus for 10 ms, then the gray re- presented for 20 ms. On those trials where the test stimulus did not appear the gray stimulus was presented for 50 ms. The period during which the gray stimulus was presented could either be preceded by or followed by presentation of the masking stimulus. The duration of the masking stimulus was variable. The following 8 durations were used : 10, 20, 40, 60, 80 , 100, 150 and 200 ms. Presentation of the mask before the gray period represented the potential for forward masking, and presentation of the mask following the gray period represented the potential for backward masking.

A total of 800 trials were presented. Each mask duration was presented on a total of 50 trials: 25 as a forward mask and 25 as a backward mask. The remaining 400 trials were catch trials where the test stimulus was not presented. Subjects were required to rate their confidence as to test stimulus presentation on a four point scale described previously. Minimum inter-trial interval was set at 200 ms in order to prevent possible carry-over masking effects from one trial to the next. Mask durations were randomised across trials, as was the temporal direction of the masking. Prior to the commencement of the experiment the intensity of the test stimulus was individually calibrated so that for the 80 ms duration mask (Backward) subjects were responing with "Yes certain" on approximately 80% of trials where the test stimulus appeared.

The next experiment was performed in the saccadic omission tradition. 20 different subjects viewed the following stimuli : The plain gray field, which was present on each trial for 50 ms, in conjunction with the mask (described above), either preceding the gray field or following it. Mask duration was variable : 0, 25, 50 or 100 ms and on any one trial, the possibilities of forward only, backward only, and forward plus backward masking existed. Where both backward and forward masks were present, each duration combination was tested, making 16 possible conditions which were randomised in presentation to subjects. The subjects' task was to rate the visibility of the gray field on a 7 point scale where 7 = perfectly visible and 0 = not at all visible. Each of the 16 conditions was presented 25 times, so each subject viewed 400 trials. The intensity of the gray field was calibrated individually for each subject so that when no masks were present, it was seen at just suprathrehold level. The method of presentation and data collection used here is to be contrasted with that used by Corfield et al (1978) in an experiment with similar aims - we are *directly* testing the masking of the gray by getting ratings for its visibility, whereas they calculated the degree of masking of the gray by getting ratings for the visibility of the masks.

For the first experiment described above, each subject's pattern of responses to test trials were compared to those of their responses to no target trials and d'e values were calculated for each mask duration using the same procedure described earlier. These values are presented in Figure 2. A statistical test was carried out to test for differences in magnitude between forward and backward masking. The results of this were that visual sensitivity under those conditions where backward masking was involved was significantly reduced compared to those conditions where forward masking was involved ($F = 6.35$, $p < .05$)

For the second experiment, mean ratings for each condition were calculated and are presented in Figure 3.

The results presented in Fig 2 clearly show that backward masking has a more powerful effect than forward masking to a target that is presented in the middle of a saccade analogue sequence. On initial inspection, this finding appears to confirm the hypothesis put forward by Breitmeyer (1980) and Matin (1974) that it is indeed backward masking that is primarily responsible as the stimulus factor producing saccadic suppression, and reject the findings of Brooks et al (1981) who found a larger forward masking effect.

We would strongly argue, however, that the results of these studies (and indeed with the addition of our results) need not be seen as incompatible.

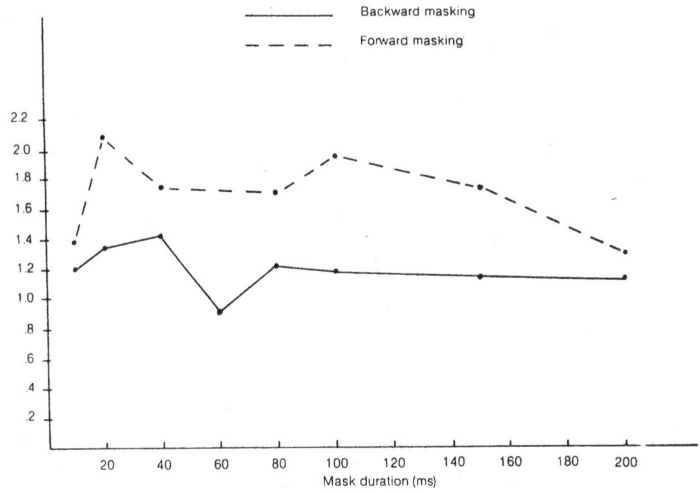

Fig.2. Mean sensitivity rating for variable duration masks (either backward or forward) in a saccade analogue using a structured target.

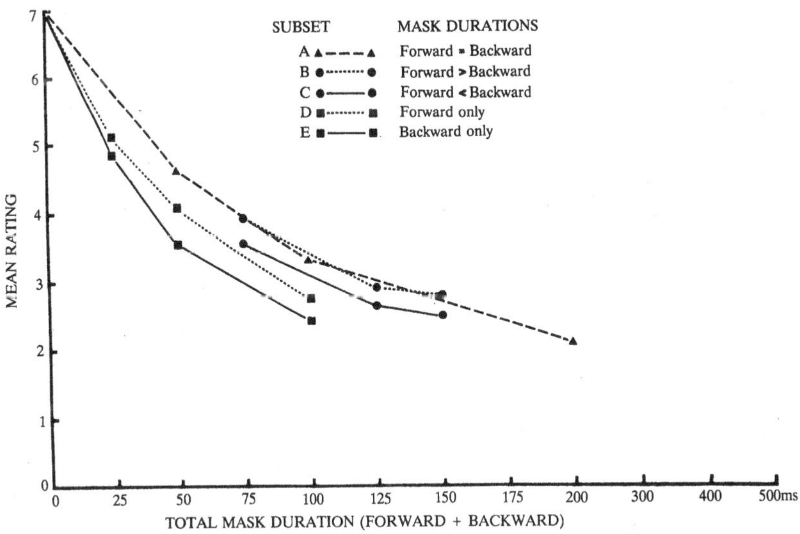

Fig. 3. Mean visibility rating for variable duration masks (backward, forward, or a combination of both) in a saccade analogue study using gray-out as target.

Firstly, the results of Brooks et al were that the forward mask is more capable of affecting stimuli placed temporally further away from it than an equivalent backward mask is capable of affecting stimuli placed temporally before it. This finding should be seen to be consistent with the whole body of literature which investigates visual persistence (e.g see Haber and Nathanson, 1968). Since stimuli are known to persist i.e are capable of having perceptual effects for some time after they have been presented, the finding that forward masking is far-reaching is not at all surprising. On the other hand, backward masking cannot rely on any perceptual persistence of the masking stimulus since the masking stimulus follows the affected test stimulus. Backward masking, therefore, does not have effects which are as far-reaching.

Such a finding is well known in the masking literature. Kietzman (1962), Schiller and Smith (1965) and Schiller (1966) have all reported that the effects of forward masking are more far-reaching than those of backward masking. Also, an examination of Figures 2 and 3 in Brooks et al (1981, pp 64-65) show that, although threshold changes in both backward and forward directions are not greatly different (no significance test were reported by the authors), there is a suggestion that the backward masking effect is stronger in that threshold was maximally raised by 3.2 log units, whereas the maximal rise in threshold for forward masking was approximately 2.8 log units.

The findings of Brooks et al then address a question which is not the equivalent to that posed as the aim of the present experiments, namely in which direction is masking greater in the saccadic situation. Given equal temporal separations of target and mask the finding of the present experiment that backward masking is more powerful is certainly not incompatible with the fact that forward maksing is longer lasting. In fact the two findings provide support for both Breitmeyer's (1980) and Matin's (1976) notion that backward masking serves in the saccadic situation to prevent the persistence of the pre-saccadic field. Two conditions need to be satisfied to support this notion. Firstly, it must be shown that the pre-sacccadic field is capable of persisting i.e. that it poses a clear threat to the perception of the post-saccadic scene. This capacity is clearly demonstated by Brooks et al (1981). Secondly, it must be shown that the effect of the post-saccade scene is a powerful one in terms of its capacity of masking the persistent pre-saccadic scene. The present experiment provides adequate evidence that this is indeed the case. The backward masking effect was observed to be strong, even at extremely short mask durations (see Fig 2)

The findings of our second experiment are more problematic. The results here show that there was no difference between backward and forward masking effects when only either form of masking was present. The expectation was, of course, that backward masking would be stronger than that of forward. One possible explanation for this result is that the omission technique used is qualitatively different from that used earlier and the results of one type of study do not necessarily need to match with the other. This would cast great doubt on the generality of the masking explanation of saccadic suppression, as presumably the masking, no matter which experimental technique is used, should have a similar effect. It is more likely, however, that a combination of a lack of statistical power and the use of naive subjects has resulted in these conflicting findings. The data plotted in Fig 3 certainly suggest that with more power available to detect effects, backward masking would indeed prove stronger than forward. In addition to this, the conditions of this experiment were recently replicated in our laboratory using a perceptually experienced observer, and a clear distinction of backward versus forward masking emerged, where backward masking clearly produced a greater effect on the visibility of the gray. This experiment also poses a further question is it duration per se, or the number of onsets and offsets which provide most effective masking. Again the results are not clear. The data in Fig 3 suggest that duration per se determines the degree of masking, whether it be forward only, backward only or backward and forward in combination. In the current replication of this experiment in our laboratory, this finding is placed under scrutiny.

It seems clear therefore that *both* backward and forward masking operate in the saccade situation. Backward masking is the more powerful of the two, but forward masking may operate over longer time spans. Stimulus based approaches to saccadic suppression should therefore now unequivocally agree that the appropriate mechanisam in operation during saccadic suppression is that of visual masking, its quantification requiring only few experiments to complete.

4. Conclusions

The present chapter has presented the case for the consideration of saccadic suppression from a functional viewpoint. In other words we have attempted to provide sufficient explanatory data to account for saccadic suppression under ordinary viewing conditions. These are those conditions of a contour-rich visual environment in photopic presentation. We have not therefore attempted to address the issue of whether there is a "residual" effect of saccadic suppression, such as that which of interest to researchers such as Volkmann (1986) and Riggs and his co-workers (e.g. see Riggs, Merton and Morton, 1974). For these researchers, questions such as whether visual sensitivity drops during eye movements made in conditions of darkness or across completely featureless fields are vital - here they are peripheral. The focus of interest here is purely on what factors might lead to a drop in visual sensitivity when saccades are made under ordinary viewing conditions. The first and most vital questions is, of course, whether when such saccades are made the mechanism responsible for the initiation of the saccade leads to sensitivity loss. A variety of techniques used to create saccade analogues have been compared to real saccades (e.g. see Brooks and Fuchs, 1975; Campbell and Wurtz, 1978; Sperling, 1990) and visual sensitivity, among other variables, measured. In all cases, the sensitivity function obtained psychophysically has mimicked that obtained with the saccade analogue. The suggestion is,therefore that the eye movement itself or the neural mechanism responsible for its initiation is not necessary to explain visual sensitivity loss during saccades. Studies at a neural level (e.g see Wurtz, 1969a; Richmond and Wurtz, 1980; Jeannerod, Kennedy and Magnin 1979) have also confirmed that when saccade analogues are compared to real saccades, responses at a neural level are similar and that a unique "corollary discharge" mechanism associated with saccades has not been found. It seems reasonable, therefore, if one is interested in functional visual behaviour to use the saccade analogue procedure to try to isolate what it is about the sequence of retinal events that leads to the visual sensitivity loss during saccades. There is general agreement that "retinal smear" can be ruled out as important in the situation (see Matin, 1974) and most researchers agree that visual masking is implicated. Unfortunately, apart from a few attempts to reduce the saccadic situation so that only particular forms of visual masking couuld potentially operate (e.g. by Matin, Clymer and Matin, 1972 and by Breitmeyer (1980) there has been no attempt to consider all possible sources of masking in the saccade situation until that of Chekaluk and Llewellyn (1990). The present chapter extends those findings to investigations in both the saccadic suppression tradition where a probe target is used and in the saccadic omission tradition where the gray-out itself is considered the effective visual stimulus. The results show clearly that pattern masking can account for visual sensitivity loss in the saccade analogue, and by implication in the true saccade made under normal conditions. This notion is not new : Matin (1974) forecast that with the amount of possible masking-type interactions occurring on the retina during the saccade it would be likely that these would indeed be responsible for much of the saccadic suppression phenomenon. The first steps to fully quantifying this prediction have finally been taken.

References

Alpern, M. Movements of the eyes. In H. Davson (ed), *The Eye* Vol. 3, New York, Academic Press, 1969.

Breitmeyer, B. G. Unmasking Visual Masking: A look at the "Why" behind the veil of the "how". *Psychological Review*, 1980, 87, 52-69.

Breitmeyer, B. G. Visual Masking: An Integrative Approach, Oxford, Clarendon Press, 1984.

Breitmeyer, B. G., & Valberg, A. Local foveal inhibitory effects of global, peripheral excitation. *Science*. 1979, 203, 463-465.

Bridgeman, B. Reply to Brooks and Fuchs: Exogenous and endogenous contributions to saccadic suppression. *Vision Research*. 1977, 17, 323-324.

Brooks, B., & Fuchs, A. The influence of stimulus parameters on visual sensitivity during saccadic eye movements. *Vision Research*. 1975, 15, 1389-1398.

Brooks, B. A., Impelman, D. M. K. & Lum, J. T. Backward and forward masking associated with saccadic eye movement. *Perception and Psychophysics*. 1981, 30(1), 62-72.

Campbell, F. W. & Kulikowski, J. J. Orientational Selectivity of the Human Visual System. *Journal of Physiology*. 1966, 187, 437-445.

Campbell, F. W. & Wurtz, R. H. Saccadic omission: Why we do not see a gray-out during a saccadic eye movement. *Vision Research*. 1978, 18, 1297-1303.

Chekaluk, E. *Saccadic Suppression, Saccadic Omission and Visual Masking*. Unpublished undergraduate thesis, University of NSW, 1980.

Chekaluk, E. & Llewellyn, K. R. Visual Stimulus Input, Saccadic Suppression, and Detection of Information from the Postsaccade Scene. *Perception and Psychophysics*. 1990, 48(2), 135-142.

Corfield, R., Frosdick, J. P. & Campbell, F. W. Grey-out elimination: The roles of spatial waveform, frequency and phase. *Vision Research*. 1978, 18, 1305-1311.

Dodge, R. Visual perception during eye movement. *Psychological Review*, 1900, 7, 454-465.

Dodge, R. The illusion of clear vision during eye movement. *Psychological Bulletin*. 1905, 2, 193-199.

Haber, R. N. & Nathanson, L. S. Post-Retinal Storage? Parks' Camel as seen through the eye of the needle. *Perception and Psychophysics*, 1968, 3, 349-355.

Jeannerod, M., Kennedy, H. & Magnin, M. Corollary discharge: Its possible implications in visual and oculomotor interactions. *Neuropsychologica*. 1979, 17, 241-258.

Kahneman, D. Method, Findings and Theory in Studies of Visual Masking. *Psychological Bulletin*, 1968, 70, 404-425.

Kietzman, M. L. The perceptual interference of successively presented visual stimuli - unpublished doctoral dissertation, University of California at Los Angeles, 1962.

Latour, P. Visual threshold during eye movements. *Vision Research*. 1962, 2, 261-262.

Lefton, L. A. Metacontrast: A Review. *Perception and Psychophysics*. 1973, 13, 161-171.

Li, W. & Matin, L. Saccadic suppression of displacement: Influence of postsaccadic exposure duration and of saccadic stimulus elimination *Vision Research*, 30, 945-955. 1990(a)

Li, W. & Matin, L. The influence of saccade length on the saccadic suppression of displacement detection *Perception and Psychophysics*, 48(5), 453-458. 1990(b)

Mackay, D. Elevation of visual threshold by displacement of retinal image. *Nature*. 225, 90-92. 1970 (a).

Mackay, D. Interocular Transfer of Suppressive Effects of Retinal Image Displacement. *Nature*. 225, 872-873. 1970(b).

Mackay, D. M. Visual Stability and Voluntary Eye Movements in Jung, R. (ed) *Handbook of Sensory Physiology*. Vol VII/3, Berlin, Springer Verlag, 1973.

Mateeff, S., Yakimoff, N. & Mitrani, L. Some characteristics of the visual masking by moving contours. *Vision Research*. 1976, 16, 489-492.

Matin, E. Saccadic suppression: A review and analysis. *Psychological Bulletin*. 1974, Vol. 81, No. 13, 899-917.

Matin, E. Saccadic Suppression and the Stable World in Monty, R. S. & Senders, J. (eds) *Eye Movements and Psychological Processes*. New Jersey, Lawrence Erlbaum, 1976.

Matin, L. Visual localization and eye movements. In K. Boff, L. Kaufman, & J. Thomas, (eds) *Handbook of perception and human performance* (Vol 1, pp 20-1 - 20-45) New York: Wiley, 1986.

Matin, E., Clymer, A. & Matin, L. Metacontrast and saccadic suppression. *Science*. 1972, 178, 179-182.

Matin, L. & Matin, E. Visual perception of direction and voluntary saccadic eye movements 1972 *Bibliotheca opthalmologica*, 82, 358-368.

Mitrani, L., Radil-Weiss, T., Yakimoff, N., Mateeff, S. & Bozkov, V. Various background pattern-effects on saccadic suppression. *Activitas Nervosa Superior*. 1975, 17, 161-163.

McMullen, M. *Is Saccadic Suppression Really Saccadic*? Unpublished undergraduate thesis, University of NSW, Sydney, 1978.

McNichol, D. *The Use of Signal Detection Theory in the Evaluation of Information Displays*. Unpublished paper presented to the NATO conference on: "Visual presentation of information", Het Vennebos, Netherlands, September 4-8, 1978.

Pearce, D. & Porter, E. Changes in visual sensitivity associated with voluntary saccades. *Psychonomic Science*. 1970, 19, 225-227.

Richards, W. Visual suppression during passive eye movement. *Journal of the Optical Society of America*. 1968, 58, 1159-1160.

Richards, W. Saccadic suppression. *Journal of the Optical society of America*. 1969, 59, 617-623.

Richmond, B. J. & Wurtz, R. H. Vision during saccadic eye movements. II. A corollary discharge to monkey superior colliculus. *Journal of Neurophysiology*. 1980, Vol. 43, No. 4, 1156-1167.

Riggs, L.A. Saccadic suppression of phosphenes: Evidence of a neural basis for saccadic suppression. In Monty, R. A. and Senders, J. W. *Eye Movements and Psychological Processes*. New York, John Wiley and Sons, 1976.

Riggs, L., Merton, P. & Morton, H. Suppression of visual phospheres during saccadic eye movements. *Vision Research*. 1974, 14, 997-1011.

Schiller, P. H. Forward and Backward Masking as a Function of Relative Overlap and Intensity of Test and Masking Stimuli. *Perception and Psychophysics*. 1966, 1, 161-164.

Schiller, P. H. & Smith, M. C. A Comparison of Forward and Backward Masking. *Psychonomic Science*. 1965, 3, 77-78.

Shebilske, W. L. Visuomotor Co-ordination in Visual Direction and Position Constancies in *Stability and Constancy in Visual Perception: Mechanisms and Processes*. (Edited by Epstein, W.) New York, Wiley, 1977.

Sperling, G. Comparison of perception in the moving and stationary eye. In E. Kowler (ed), *Eye Movement and their Role in Visual and Cognitive Processes*. Amsterdam, The Netherlands: Elsevier Biomedical Press, 1990. Pp. 307-351.

Stark, L., Kong, R., Schwartz, S., Hendry, D. & Bridgeman, B. Saccadic suppression of image displacement *Vision Research* (1976), 16, 1185-1187.

Stoper, A. E. & Mansfield, J. G. Metacontrast and parcontrast suppression of a contourless area. *Vision Research*. 1978, 18, 1669-1674.

Uttal, W. & Smith, P. Recognition of alphabetic characters during voluntary eye movements. *Perception and Psychophysics*. 1968, 3, 257-264.

Volkmann, F. C. Vision during voluntary saccadic eye movements. *Journal of the Optical Society of America*. 1962, 52, 571-578.

Volkmann, F. C. Human visual suppression *Vision Research*. (1986) 26(9), 1401-1416.

Volkmann, F., Schick, A. & Riggs, L. Time course of visual inhibition during saccades. *Journal of the Optical Society of America*. 1968, 58, 562-569.

Woodworth, R. Vision and localization during eye movements. *Psycholgical Bulletin*. 1906, 3, 68-70. (Abstract)

Wurtz, R. Response to striate cortex neurons to stimuli during rapid eye movements in the monkey. *Journal of Neurophysiology*. 1969, 32, 975-986. (a)

Wurtz, R. Comparison of the effects of eye movements and stimulus movements of striate cortex neurons on the monkey. *Journal of Neurophysiology*. 1969, 32, 987-994. (b)

Wurtz, R. H., Richmond, B. J. & Judge, S. J. Vision during saccadic eye movements. III. Visual Interactions in monkey superior colliculus. *Journal of Neurophysiology*. 1980, Vol. 43, No. 4, 1168-1181.

Zaretsky, M. Quantitative measurements of centrally and retinally generated saccadic suppression in a locust movement detector neurone. *Journal of Physiology*. 1982, 328, 521-533.

Zuber, B. L., Crider, A. & Stark, L. Saccadic suppression associated with microsaccades. *QPR 74*. Research Laboratory of Electronics, MIT, 1964, 244-249. From Zuber and Stark (1966).

Zuber, B. & Stark, L. Saccadic suppression: elevation of visual threshold associated with saccadic eye movements. *Experimental Neurology*. 1966, 16, 65-79.

The Role of Eye Movements in Perceptual Processes
E. Chekaluk and K.R. Llewellyn (Editors)

MECHANISMS UNDERLYING SACCADIC SUPPRESSION IN SPECIFICALLY DISABLED AND NORMAL READERS

WILLIAM LOVEGROVE

1. Introduction

The research to be described has been concerned with the mechanisms argued to be involved in saccadic suppression in a particular group of children who have trouble learning to read. Children in this group are commonly referred to as specifically reading disabled (SRDs) and are defined as children of normal intelligence, with normal education opportunities, no brain damage, and no gross behavioural problems who nevertheless read at least two years behind the level expected for this age and intelligence.

For some years the commonly accepted view within the reading disability literature has been that reading disability is not attributable to visual deficits and that normal and specifically-disabled readers (SRDs) do not differ systematically in terms of visual processing (Benton, 1962; Vellutino, 1979a, 1979b). Extensive work over the last ten years, however, has clearly demonstrated that the two groups do differ in terms of visual processing.

This has been brought about partially by developments in theoretical vision which have been applied to reading thus, providing a more meaningful theoretical context in which to consider reading disability and vision. The following section outlines one approach to vision which has been usefully applied.

2. Spatial Frequency Processing

One approach to vision research (Campbell, 1974; Graham, 1980) indicates that information is transmitted from the retina to the brain via a number of separate parallel pathways. The separate pathways are frequently referred to as channels. Each channel is specialised to process information about particular features of visual stimuli.

The properties of channels often have been investigated using patterns like those shown in Figure 1. These patterns are usually called sine-wave gratings. Two properties of these patterns are of interest in this chapter:

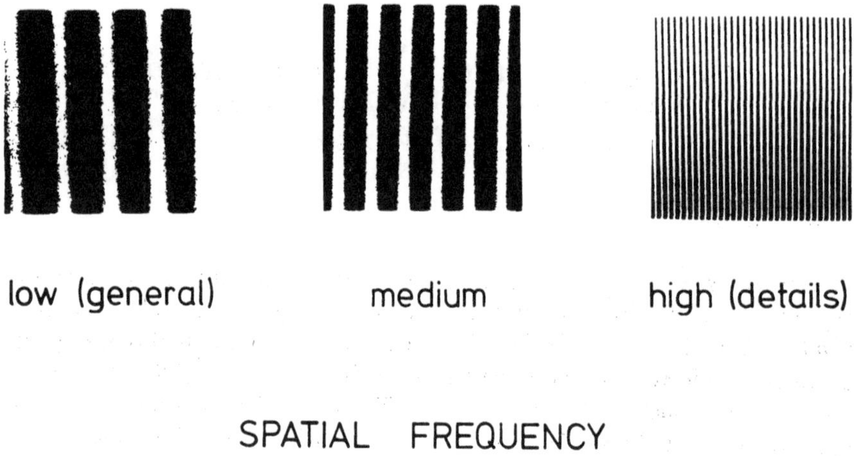

Fig. 1. Sine-wave gratings commonly used in vision research concerned with spatial frequency channels. Low spatial frequencies are shown on the left and high spatial frequencies on the right.

1. Spatial frequency, which refers to the number of cycles (one dark plus one light bar) per degree of visual angle (c/deg). Spatial frequency is higher on the right than the left in Figure 1. Spatial frequency can be thought of in terms of stimulus size

2. Contrast, which refers to the difference between the maximum and minimum luminances of the grating. It is a measure of the ratio of the brightest to the darkest section of the pattern.

Research (Campbell, 1974; Graham, 1980) has identified a number of channels each sensitive to a narrow range of spatial frequencies (or stimulus widths) and orientations in cats, monkeys and humans.

Spatial frequency or size-sensitive channels are relevant to reading because when we read we process both general (low spatial frequency) and detailed (high spatial frequency) information in each fixation. We extract detailed information from an area approximately 5 - 6 letter spaces to the right of fixation. Beyond this we also extract visual information but only of a general nature such as word shape (Rayner, 1975). These two types of size information must be combined in some way.

It has also been shown that the different channels transmit their information at different rates and respond differently to different rates of temporal change. Some channels are sensitive to very rapidly changing stimuli and others to stationary or slowly moving stimuli. Such results have led to the proposal of two subsystems within the visual system. This division is believed to be important in combining the two types of size information involved in reading.

3. The Sustained and Transient Subsystems

It has been shown that spatial frequency channels differ in their temporal properties. In a typical experiment subjects are shown patterns like those in Figure 1 flickering at various rates.

Subjects are required to set contrast levels so that they just can see either flicker or pattern. When low spatial frequency gratings are moving quickly, we see flicker at lower contrasts than we see pattern but we experience the reverse at high spatial frequencies. Separate measures can be taken of our sensitivity to flicker and pattern with a range of different sized stimuli (spatial frequencies) flickering at different speeds. Thus we can plot sensitivity functions for pattern and flicker thresholds at a range of spatial frequencies. With large stimuli (low spatial frequencies) we are more sensitive to rapidly changing stimuli but with small stimuli (high spatial frequencies) we are more sensitive to stationary or slowly moving stimuli. The two functions obtained from such experiments are believed to measure two subsystems in the visual system, the transient and sustained subsystems. An extensive discussion of the properties of these systems and how they are identified can be found in Breitmeyer (1988). Breitmeyer also discusses the evidence indicating the physiological basis of these two systems.

The properties of these two subsystems have been identified and are shown in Table 1.

Table 1. General Properties of the Sustained and Transient Subsystems

Sustained System	Transient System
Less sensitive to contrast	Highly sensitive to contrast
Most sensitive to high spatial frequencies	Most sensitive to low spatial frequencies
Most sensitive to low temporal frequencies	Most sensitive to high temporal frequencies
Slow transmission times	Fast transmission times
Responds throughout stimulus presentation	Responds at stimulus onset and offset
Predominates in central vision	Predominates in peripheral vision
The sustained system may inhibit the transient system	The transient system may inhibit the sustained system

It has been demonstrated physiologically (Singer & Bedworth, 1973) and psychophysically that the two systems may inhibit each other (Breitmeyer & Ganz, 1976). In particular if the sustained system is responding when the transient system is stimulated, the transient will terminate the sustained activity. An example of how this may occur is as follows. If we are fixating on the detail of an object and a stimulus moves into the periphery of our vision, the transient system is likely to inhibit or override the sustained system until we know what is in our peripheral vision. How this may have evolved is easier to imagine if we consider not a human reading but a rabbit eating and a predator appearing to the side. There would be survival value for the rabbit in having the transient system inhibit the sustained system until the level of threat could be determined. These two subsystems and the interactions between them may serve a number of functions essential to the reading process.

The transient system is predominantly a flicker or motion detecting system which transmits information about stimulus change and general shape. The spatial information it transmits is coarse and thus well suited for transmitting peripheral information in reading. The sustained system is predominantly a detailed pattern detecting system which transmits information about stationary stimuli. In reading the sustained system should be most important in extracting detailed information during fixations and the transient system in extracting general information from the periphery. Below we shall see that the two systems also interact in important ways.

4. Sustained and Transient Subsystems and Reading

When reading, the eyes move through a series of rapid eye movements called saccades. These are separated by fixation intervals when the eyes are stationary. Saccadic eye movements are generally in the direction of reading, that is, from left to right when reading English. Sometimes the eyes also move from right to left in what are called regressive eye movements or regressions. The average fixation duration is approximately 200 - 250 msec for normal readers and it is during these stationary periods that information from the printed page is seen. The average saccade length is 6 - 8 characters or about 2 degrees of visual angle (Rayner & McConkie, 1976). Saccadic eye movements function to bring unidentified regions of text into foveal vision for detailed analysis during fixations. Foveal vision is the area of high acuity in the centre of vision extending approximately 2 degrees (6 - 8 letters) around the fixation point on a line of text. Beyond the fovea acuity drops off dramatically.

The role of transient and sustained subsystems in reading has been considered by Breitmeyer (Breitmeyer 1980, 1983; Breitmeyer & Ganz, 1976). Figure 2 represents the hypothesised activity in the transient and

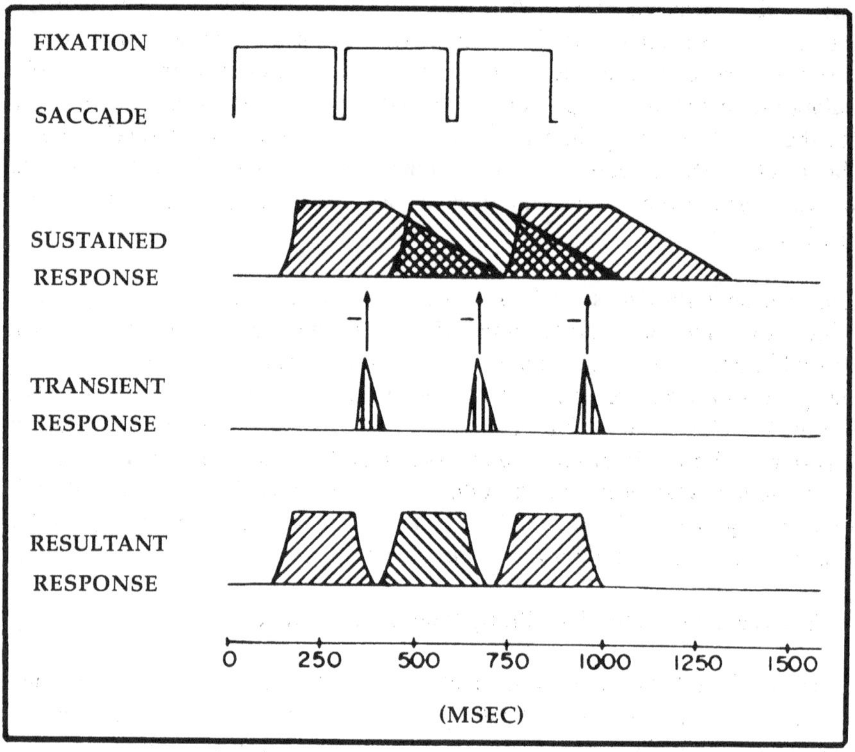

Fig. 2. A hypothetical response sequence of sustained and transient channels during three 250 msec fixation intervals separated by 25 msec saccades (panel 1). Panel 2 illustrates persistence of sustained channels acting as a forward mask from preceding to succeeding fixation intervals. Panel 3 shows the activation of transient channels shortly after each saccade which exerts inhibition (arrows with minus signs) on the trailing, persisting sustained activity generated in prior fixation intervals. Panel 4 shows the resultant sustained channel response after the effects of the transient-on-sustained inhibition have been taken into account (From "Unmasking Visual Masking : A look at the 'Why' behind the veil of the 'How'" by B.G. Breitmeyer, *Psychological Review*, 1980, *82*, 52-69. Copyright 1980 by the American Psychological Association. Permission to reprint granted.)

sustained channels over a sequence of 3 fixations of 250 msec duration separated by 2 saccades of 25 msec duration each.

The sustained channel response occurs during fixations and may last for several hundred milliseconds. This response provides details of what the eye is seeing. The transient channel response is initiated by eye movements and lasts for much shorter durations. Consequently both systems are involved in reading. The duration of the sustained response may outlast the physical duration of the stimulus. This is a form of visual persistence produced by the activation of sustained channels. Its duration increases with increasing spatial frequency (Bowling, Lovegrove & Mapperson, 1979; Meyer & Maguire, 1977) and may last longer than a saccade.

If sustained activity (as shown in Figure 2, panel 2) generated in a preceding fixation persists into the succeeding one, it may interfere with processing there. In this case, what may happen when reading a line of print requiring one, two or three fixations is illustrated in Figure 3 (adapted from Hochberg, 1978).

Consequently it is evident that for tasks such as reading, persistence across saccades presents a problem as it may lead to superimposition of successive inputs. Breitmeyer proposes that the problem posed by visual persistence is solved by rapid saccades as shown in the bottom two panels of Figure 2. Saccades not only change visual fixations, they also activate short latency transient channels (panel 3) which are very sensitive to stimulus movement.

These transient channels, in turn, inhibit the sustained activity which would persist from a preceding fixation and interfere with the succeeding one (Breitmeyer & Ganz, 1976; Matin, 1974). The result is a series of clear, unmasked and temporally segregated frames of sustained activity, each one representing the pattern information contained in a single fixation as shown in panel 4 of Figure 2.

In these terms, clear vision on each fixation results from interactions between the sustained and transient channels. Consequently the nature of transient-sustained channel interaction seems to be important in facilitating normal reading. Any problem in either the transient or the sustained system or in the way they interact may have harmful consequences for reading.

5. Transient and Sustained Processing in SRDs and Controls

The possibility of a visual deficit in SRDs has recently been investigated within a spatial frequency analysis framework. The following is not a complete review of all recent research but a summary of the research carried

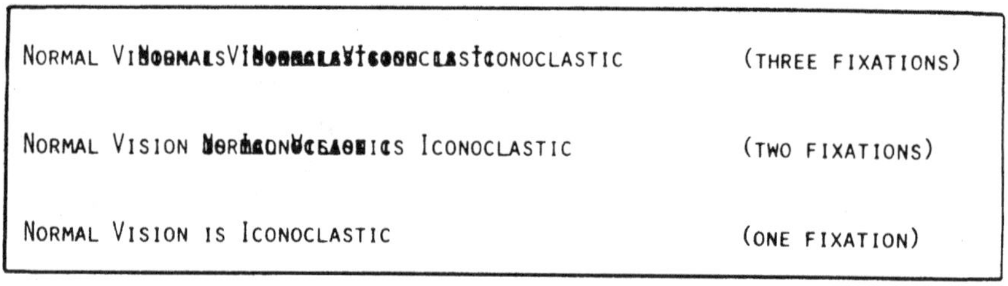

Fig. 3. The perceptual masking effects of temporal integration of persisting sustained activity from preceding fixation intervals with sustained activity generated in succeeding ones when the reading of a printed sentence requires one, two or three fixations. Here, as in panel 2 of Fig. 2, the effects of transient-on-sustained inhibition have not been taken into account. (From "Unmasking Visual Masking : A Look at the 'Why' behind the Veil of the 'How'" by B.G. Breitmeyer, *Psychological Review*, 1980, *82*. 52-69. Copyright 1980 by the American Psychological Association. Permission to reprint granted.)

out in a few laboratories including ours. Much of this research has been directed at the functioning of the transient and sustained systems in normal and specifically-disabled readers.

6. Low-Level Visual Processing in Controls and SRDs

Visible persistence is one measure of temporal processing in spatial frequency channels and refers to the continued perception of a stimulus after it has been physically removed. The top panel in Figure 2 demonstrates such persistence. This is assumed to reflect ongoing neural activity initiated by the stimulus presentation. In adults, duration of visible persistence increases with increasing spatial frequency (Bowling et al., 1979; Meyer & Maguire, 1977). Several studies have compared SRDs and controls on measures of visible persistence. It has been shown that SRDs aged from 8 to 15 years have a significantly smaller increase in persistence duration with increasing spatial frequency than do controls (Badcock & Lovegrove, 1981; Lovegrove, Heddle & Slaghuis, 1980; Slaghuis & Lovegrove, 1985).

When visible persistence is measured in both groups under conditions which reduce transient system activity (using a uniform field flicker mask), persistent differences between the groups essentially disappear (Slaghuis & Lovegrove, 1984). This finding suggests that SRDs may differ from controls mainly in the functioning of their transient systems.

The two groups have also been compared on a task which measures the minimum contrast (refer to Figure 1) required to see a pattern. Contrast sensitivity (the reciprocal of the minimum contrast required for detection), plotted as a function of spatial frequency, is referred to as the contrast sensitivity function (CSF). Pattern CSFs have been measured in at least 5 separate samples of SRDs and control readers with ages ranging from 8 years to 14 years. It has generally been shown that SRDs are less sensitive than controls at low (1.0 to 4 c/deg) spatial frequencies (Lovegrove, Bowling, Badcock & Blackwood, 1980; Lovegrove et al., 1982; Martin & Lovegrove, 1984). In some studies the two groups do not differ in contrast sensitivity at higher (12 to 15 c/deg) spatial frequencies (Lovegrove et al., 1980) and in others SRDs are slightly more sensitive than controls in that range (Lovegrove et al., 1982; Martin & Lovegrove, 1984). At high luminances SRDs have been found to be less sensitive at high spatial frequencies than controls. Once again inhibition of the transient system by uniform-field masking influenced the SRDs less than controls (Martin & Lovegrove, 1988) thus supporting the notion of a difference between the groups in the transient system.

A third line of research has measured transient system functioning more directly than did the previous two measures. It has been argued that flicker thresholds are primarily mediated by the transient system. Consequently flicker thresholds under a range of conditions have been measured in SRDs and controls. In these experiments subjects are shown a sine-wave grating counterphasing, i.e., moving from right to left and back the distance of one cycle at whatever speed the experimenter chooses. Subjects are required to detect the presence of the flicker. In a number of experiments SRDs have been shown to be less sensitive than controls to counterphase flicker (Brannan & Williams, 1988; Martin & Lovegrove, 1987, 1988). The differences between the groups sometimes become larger as the temporal frequency increases (Martin & Lovegrove, 1987, 1988) and sometimes does not (Brannan & Williams, 1988). What happens depends on the spatial make up of the stimuli. This is a direct measure of transient system processing and distinguishes very well between individuals in the two groups (Martin & Lovegrove, 1987).

Additional support for differences between the groups in terms of spatial frequency processing comes from a recent visual evoked potential study (May, Lovegrove, Martin & Nelson, 1991; May, Dunlap & Lovegrove, 1991). Subjects were presented with sine-wave gratings ranging in spatial frequency from 0.5 to 8.0 cycles per degree flickering at a rate of two Hertz (Hz). Stimulus duration was 200msec. This allowed analysis of two components of the VEP elicited by both stimulus onset and by stimulus offset. The major findings indicated that poor readers had significantly lower amplitudes and significantly shorter latencies for components produced by stimulus offsets when low spatial frequency stimuli were used. Further analyses of these data revealed two factors for both the low and high spatial frequency stimuli (May, Dunlap and Lovegove, 1991). With the low spatial frequency stimulus, Factor II was associated with the latencies on the first onset component and Factor I with the latencies of all components. These scores were subject to a discriminant analysis which showed that good and poor readers were well differentiated by the factor scores on the low spatial frequency but not the high spatial frequency factor. This is consistent with a problem in the transient system. Further visual-evoked potential data supporting this conclusion but using different conditions, have been reported (Livingstone, Drislane & Galaburda, 1991).

Lovegrove and associates have also conducted a series of experiments comparing sustained system processing in controls and SRDs (see Lovegrove et al., 1986). Using similar procedures, equipment and subjects as the experiments outlined above, this series of experiments has failed to show any significant differences between the two groups. This implies that either there are no differences between the groups in the functioning of their sustained systems or that such differences are smaller than the transient differences demonstrated.

In summary, four converging lines of evidence suggest a transient deficit in SRDs. The differences between the groups are quite large on some measures and discriminate well between individuals in the different groups with approximately 75 percent of SRDs showing reduced transient system sensitivity (Slaghuis & Lovegrove, 1985). At the same time evidence to date suggests that the two groups do not differ in sustained system functioning.

7. Higher Level Perceptual Processes and SRD

It is known that the transient and sustained systems may be involved in higher-level perceptual processes than those discussed above (Breitmeyer & Ganz, 1976; Weisstein, Ozoz & Szoc, 1975). Williams and colleagues have recently investigated the question of how a transient deficit may manifest itself in a range of higher-level perceptual processes. Their general conclusion is that SRDs manifest difficulties on a large number of perceptual tasks, most of which are believed to involve the transient system (Williams & LeCluyse, 1990).

In an important study Williams, Molinet and LeCluyse (1989) plotted the time course of transient-sustained interactions. A standard way of measuring the temporal properties of transient-sustained interactions is to use a metacontrast masking paradigm (Breitmeyer & Ganz, 1976). In metacontrast masking a target is briefly presented followed at various delays by a spatially adjacent masking stimulus. The experiment measures the effect of the mask on the visibility of the target. The target is affected by both the temporal and spatial relationship between the mask and the target. It is normally found that the visibility of the target first decreases and then increases as the pattern mask follows it by longer and longer delays. Breitmeyer and Weisstein have argued that metacontrast masking is due to the inhibition of the sustained response to the target by the transient response to the mask. This happens in much the same way sustained persistence is terminated by transient activity during reading as is shown in Figure 2.

Maximal masking occurs when the transient response to the mask and the sustained response to the target overlap most in time in the visual system.

This occurs in metacontrast when the mask follows the target by a certain interval. The point of maximal masking, then, provides an index of the relative processing rates of the target sustained response and the mask transient response. If the difference in rate of transmission is small, the dip in the masking function occurs after a short delay and vice versa. The magnitude of masking provides an index of the strength of transient-on-sustained inhibition. Additionally metacontrast is normally stronger in peripheral than in central viewing presumably because of the preponderance of transient pathways in peripheral vision. It should be noted that this is the same mechanism proposed to be involved in saccadic suppression (the suppression of sustained activity during eyemovements) as discussed earlier in relation to Figures 2 and 3.

In an experiment using line targets Williams, Molinet and LeCluyse (1989) showed that maximal masking occurred at a shorter delay in SRDs than in controls. This result is direct evidence that SRDs have a slower transient system or at least a smaller difference between the rates of processing for their transient and sustained systems than controls. They also found that in peripheral vision SRDs experienced almost no metacontrast masking which further supports this position. The magnitude of masking was also less in central vision showing that the transient inhibition was also weaker. Further evidence supporting timing differences between the transient and sustained systems in controls and SRDs has been provided in a subsequent metacontrast experiment where subjects had to identify a target letter. A letter mask was used but it could be combined with the target to form either a word or a nonword (Williams, Brannan & Bologna, 1988). Both of these studies provide clear evidence of temporal differences between the two groups contributing to high-level perceptual problems.

In summary there are now a large number of studies which have investigated higher-level perceptual processing in good and poor readers. The results from this wide range of measures confirm the finding of a transient deficit in SRDs. They also suggest that there may be other deficits (visual and higher level) but the precise nature of these is not yet clear.

8. What About the Confusion in the Literature?

Over the years a substantial number of studies have reported differences in visual processing between good and poor readers. Several researchers (Blackwell, McIntyre, & Murray, 1983; Di Lollo, Hanson, & McIntyre, 1983; Hoien, 1980; Lovegrove & Brown, 1978; Stanley & Hall, 1973) have shown that masking occurs over longer durations in SRDs than in controls. Mason,

Pilkington and Brandau (1981) have shown SRDs to have difficulties with order rather than item information. Hyvarinen and Laurinen (1980) have measured spatial and temporal processing across spatial frequencies. They generally found that disabled readers were less sensitive than controls without specifying whether this difference was greater at certain spatial frequencies.

The difficulty in making sense of this literature, however, is that for almost every study showing differences between the two groups another study may be cited failing to show differences. For example Arnett and Di Lollo (1979), Fisher and Frankfurter (1977), Morrison, Giordani and Nagy (1977) and Manis and Morrison (1982) have all conducted studies with short-duration stimuli without finding any significant differences between groups. Howell, Smith and Stanley (1981) and Smith, Early and Grogan (1986) failed to show spatial-frequency specific differences in visible persistence between the two groups.

An obvious question is whether it is possible to reconcile these different sets of results in terms of the argument presented here. In the context of this chapter it may be suggested that many of the studies which have failed to show differences between dyslexics and normal readers in visual processing may have measured sustained processing and those which have shown differences have measured either transient system processing or transient-sustained interactions. Support for this position has recently been provided by Meca (1985) who has conducted a meta-analysis on a large number of studies investigating vision and reading. He plotted effect size as a function of spatial frequency. As would be expected if SRDs had a transient system problem but not a sustained system problem, effect size was greatest at low spatial frequencies and decreased with increasing spatial frequency (Meca, 1985).

While this is almost certainly over simplistic it does allow us to make predictions about what should be found on a range of different tasks depending on whether or not transient or sustained processing is being measured. A recent study (Solman and May, 1990) has investigated spatial localisation in dyslexic and normal readers within this context. They predicted not only conditions where dyslexics should be worse than controls but also where dyslexics would perform at least as well or even better than controls. They found that when the targets were close to the fixation points SRDs performed slightly better than controls. This pattern reversed as the targets moved more into peripheral vision (and were presumably processed more by the transient system). It may be stated, therefore, that while there is a large amount of data consistent with the argument presented here there are also substantial data inconsistent with it. The consistent data are, generally, more recent and have formed part of one of a small number of systematic programs of research. If the argument presented here is valid, it is possible to make clear predictions about the types of visual tasks on which SRDs should do worse and/or better than controls.

It is important to note in this context that recent research has not simply demonstrated SRDs performing more poorly than controls on all measures of visual processing. SRDs have been shown to perform at least as well as or even better than controls on some tasks, e.g., high spatial frequency sensitivity, visual acuity and the oblique effect (Lovegrove et al., 1986). Generally this is thought to be the case on tasks measuring sustained system functioning. Further experimentation will determine whether or not this is so.

9. Possible Relationship to Other Problems Manifest in SRDs?

There are two issues to be addressed by any theory of specific reading disability which attributes a role to visual factors. The first concerns the relationship between visual deficits and other known processing deficits manifested by SRDs especially those in phonological awareness (Bradley & Bryant, 1983) and working memory (Jorm, 1983). The second is to do with whether a visual approach is able to predict conditions under which normal readers and SRDs may perform differently. Both issues will be briefly discussed.

There is an extensive evidence that SRDs perform worse than controls in a number of other areas, especially in aspects of phonological awareness and working memory. It becomes important to ask what, if any, is the relation between the transient system deficits and these other processing areas.

The issue was considered in a recent study of approximately sixty SRDs and sixty controls (Lovegove, McNicol, Martin, Mackenzie & Pepper, 1988). They took measures of transient system processing, phonological recoding and working memory in each child. These measures were subjected to a factor analysis which showed that some of the phonological recoding measures loaded on the same factor as the measures of transient processing used. This shows some relation between the two processing areas but, of course, does not reveal the precise nature of that relationship. Until this relationship is further clarified it is premature to reject the possibility of a link between visual and phonological processes in reading. The measures of working memory used did not load on the same factor as did the transient system measures and the phonological recoding measures. This study thus provides some preliminary evidence of a link between phonological recoding and visual processing in SRDs but the exact nature of this relation is still to be determined.

It is possible that some of these different deficits are related by virtue of the fact that some SRDs have a problem in processing rapidly presented stimuli in all sensory modalities. Tallal (1980), for example, has shown similar problems in audition as we have shown in vision. Livingstone, Drislane and Galaburda (1991) have recently noted that the auditory and somatosensory systems may also be subdivided into fast and slow components like the visual system. They then speculated that problems in each fast system may occur in SRDs. Even though this possibility has not yet been directly investigated, it is an exciting prospect which may help to integrate a large amount of apparently discrepant data.

The second issue raised above concerns the possibility of predicting conditions which would lead to different levels of performance in the two groups. In terms of Breitmeyer's theory outlined earlier a transient deficit should lead to more errors for SRDs when reading continuous text than when reading isolated words. This is because this task requires integration of peripheral information from one fixation with central information on the next. This has recently been tested by varying the mode of visual presentation. Three conditions of visual presentation on a computer monitor were used while holding the semantic context constant. This was done by presenting stories in three different ways. In the first condition one word at a time was presented in the middle of the screen. Thus the subjects never had to move their eyes and never had information presented to the right of fixation. In the second condition one word was presented at a time but its position was moved across the screen. Here the subjects were required to move their eyes across the screen but still were never presented with information to the right of fixation. The final condition was a whole line presentation which most closely approximated normal reading. Rate of word presentation was held constant across the three conditions. The results (Lovegrove & MacFarlane, 1990) showed normal readers were most accurate in the whole line condition and made more errors in the two one-word at time conditions. The reverse was true for the SRDs. They read significantly more accurately in both one-word conditions than the whole-line condition.

The mode of presentation of written material which maximised reading accuracy in controls, therefore, produced the most errors in SRDs. These findings have both theoretical and practical implications.

10. Conclusions

The data reported in the last ten years show that many SRDs have a particular visual deficit. It has also been shown that it is unlikely that the transient deficit results from being unable to read (Lovegrove et al., 1986) although it is not yet known how it may contribute to reading difficulties. This problem appears to be present in a large percentage of disabled readers and not just for a sub-group frequently referred to as visuo-spatial dyslexics. There is still a lot of work to be done before knowing how this processing difficulty relates to other difficulties. The results with different modes of visual presentation will be further investigated in our laboratory and are encouraging.

References

Arnett, J.L., & Di Lollo, V. (1979). Visual information processing in relation to age and reading ability. *Journal of Experimental Child Psychology, 27*, 143-152.

Badcock, D.R., & Lovegrove, W. (1981). The effect of contrast, stimulus duration and spatial frequency on visible persistence in normal and specifically disabled readers. *Journal of Experimental Psychology; Human Perception and Performance, 7*, (3), 495-505.

Benton, A.L. (1962). Dyslexia in relation to form perception and directional sense. In J. Money (Ed.), *Reading disability; Progress and research needs in dyslexia*, (pp 81-102), Johns Hopkins Press.

Blackwell, S., McIntyre, D., & Murray, M. (1983). Information processing from brief visual displays by learning disabled boys. *Child Development, 54*, 927-940.

Bowling, A., Lovegrove, W., & Mapperson, B. (1979). The effect of spatial frequency and contrast on visible persistence. *Perception, 8*, 529-539.

Bradley, L., & Bryant, P. (1983). Categorising sounds and learning to read - a causal connection. *Nature, 301*, 419-421.

Brannan, J., & Williams, M. (1987). Allocation of visual attention in good and poor readers. *Perception and Psychophysics, 41*, 23-28.

Brannan, J., & Williams, M. (1988). The effects of age and reading ability on flicker threshold. *Clinical Visual Sciences, 3*, 137-142.

Breitmeyer, B.G. (1988). Reality and relevance of sustained and transient channels in reading and reading disability. *Paper presented to the 24th International congress of Psychology*, Sydney.

Breitmeyer, B.G. (1980). Unmasking visual masking: A look at the "why" behind the veil of "how". *Psychological Review, 87*, (1), 52-69.

Breitmeyer, B.G. (1983). Sensory masking, persistence and enhancement in visual exploration and reading. In K. Rayner (Ed.). *Eye movements in reading: Perceptual and language processes*. New York: Academic Press.

Breitmeyer, B.G., & Ganz, L. (1976). Implications of sustained and transient channels for theories of visual pattern making, saccadic suppression and information processing. *Psychological Review, 83*, 1-36.

Campbell, F.W. (1974). The transmission of spatial information through the visual system. In F.O. Schmidt & F.S. Worden (Eds.). *The Neurosciences Third Study Program*, (pp 95-103), Cambridge, Massachusetts: The M.I.T. Press.

Di Lollo, V., Hanson, D., & McIntyre, J. (1983). Initial stages of visual information processing in dyslexia. *Journal of Experimental Psychology: Human Perception and Performance, 9*, 923-935.

Fisher, D.F., & Frankfurter, A. (1977). Normal and disabled readers can locate and identify letters: Where's the perceptual deficit? *Journal of Reading Behaviour, 10*, 31-43.

Galaburda, A., Drislane, F. & Livingstone, M. (1991). Anatomical evidence for a magnocellular defect in developmental dyslexia. *Proceedings of the New York Academy of Science*, (in press).

Graham, N. (1980). Spatial frequency channels in human vision. Detecting edges without edges detectors. In C.S. Harris (Ed.), *Visual coding and adaptability*. (pp 215-262), Hillsdale: Lawrence Erlbaum Associates Inc.

Hochberg, J.E. (1978). *Perception*. Englewood Cliffs, New Jersey: Prentice-Hall.

Hoien, T. (1980). The relationship between iconic persistence and reading disabilities. In Y. Zotterman (Ed.). *Dyslexia: Neuronal, cognitive and linguistic aspects*. (pp 93- 107), Oxford: Pergamon Press.

Howell, E.R., Smith, G.A., & Stanley, G. (1981). Reading disability and visual spatial frequency specific effects. *Australian Journal of Psychology, 33*, (1), 97-102.

Hyvarinen, L., & Laurinen, P. (1980). Ophthalmological findings and contrast sensitivity in children with reading difficulties. In Y. Zotternman (Ed.), *Dyslexia: Neural, cognitive and linguistic aspects*. (pp 117-119), Oxford: Pergamon Press.

Jorm, A. (1983). Specific reading retardation and working memory: a review. *British Journal of Psychology, 74*, 311-342.

Livingstone, M., Drislane, F. & Galaburda, A. (1991). Physiological evidence for a magnocellular defect in developmental dyslexia. *Proceedings of the New York Academy of Science*, (in press).

Lovegrove, W.J., Bowling, A., Badcock, D., & Blackwood, M. (1980). Specific reading disability: Differences in contrast sensitivity as a function of spatial frequency. *Science, 210*, 439-440.

Lovegrove, W.J., & Brown, C. (1978). Development of information processing in normal and disabled readers. *Perceptual and Motor Skills, 46*, 1047-1054.

Lovegrove, W., Heddle, M., & Slaghuis, W. (1980). Reading disability: Spatial frequency specific deficits in visual information store. *Neuropsychologia, 18*, 111-115.

Lovegrove, W., Martin, F., Eowling, A., Badcock, D., & Paxton, S. (1982). Contrast sensitivity functions and specific reading disability. *Neuropsychologia, 20*, 309-315.

Lovegrove, W., Martin, F., & Slaghuis, W. (1986). A theoretical and experimental case for a residual deficit in specific reading disability. *Cognitive Neuropsychology, 3*, 225-267.

Lovegrove, W., McNicol, D., Martin, F., Mackenzie, B., & Pepper, K. (1988). Phonological recoding, memory processing and memory deficits in specific reading disability. In D. Vickers and P. Smith (Eds.). *Human Information Processing: Measures, Mechanisms and Models*, North-Holland: Amsterdam, 65-82.

Manis, F.R., & Morrison, F.J. (1982). Processing of identity and position information in normal and disabled readers. *Journal of Experimental Child Psychology, 33*, (1), 74-86.

Martin, F., & Lovegrove, W. (1984). The effects of field size and luminance on contrast sensitivity differences between specifically reading disabled and normal children. *Neuropsychologia, 22*, 75-77.

Martin, F., & Lovegrove, W. (1987). Flicker contrast sensitivity in normal and specifically-disabled readers. *Perception, 16*, 215-221.

Martin, F., & Lovegrove, W. (1988). Uniform & field flicker in control and specifically-disabled readers. *Perception, 17*, 203-214.

Mason, M., Pilkington, C., & Brandau, R. (1981). From print to sound: Reading ability and order information. *Journal of Experimental Psychology: Human Perception and Performance, 7*, 580-591.

Matin, E. (1974). Saccadic suppression: A review and an analysis. *Psychological Bulletin, 81*, 899-915.

May, J., Dunlap, W. & Lovegrove, W. (1991). Visual evoked potentials latency factor scores differentiate good and poor readers. *Clinical Vision Sciences*. In Press.

May, J., Lovegrove, W., Martin, F., & Nelson, W. (1991). Pattern-elicited visual evoked potentials in good and poor readers. *Clinical Vision Sciences, 2*, 131-136.

Meca, J. (1985). La Hipotesis del deficit perceptivo del retraso especifico en lectura: un estudio meta-analitico. *Anales de Psicologia, 2*, 75-91.

Meyer, G.E. & Maguire, W.M. (1977). Spatial frequency and the mediation of short-term visual storage. *Science, 198*, 524-525.

Morrison, F., Giordani, B., & Nagy, J. (1977). Reading disability: An information processing analysis. *Science, 196*, 77-79.

Rayner, K. (1975). The perceptual span and peripheral cues in reading. *Cognitive Psychology, 7*, 65-81.

Rayner, K., & McConkie, G.W. (1976). What guides a reader's eye movements? *Vision Research, 16*, 829-837.

Singer, W., & Bedworth, N. (1973). Inhibitory interaction between X and Y units in the cat lateral geniculate nucleus. *Brain Research, 49*, 291-307.

Slaghuis, W., & Lovegrove, W.J. (1984). Flicker masking of spatial frequency dependent visible persistence and specific reading disability. *Perception, 13*, 527-534.

Slaghuis, W., & Lovegrove, W.J. (1985). Spatial-frequency mediated visible persistence and specific reading disability. *Brain and Cognition, 4*, 219-240.

Smith, A., Early, F., & Grogan, S. (1986). Flicker masking and developmental dyslexia. *Perception, 15*, 473-482.

Solman, R. & May, J. (1990). Spatial localisation discrepancies: a visual deficit in reading. *American Journal of Psychology, 103*, 243-263.

Stanley, G., & Hall, R. (1973). Short-term visual information processing in dyslexics. *Child Development, 44*, 841-844.

Stein, J., & Fowler, S. (1982). Diagnosis of dyslexia by a new indicator of eye dominance. *British Journal of Ophthalmology, 66*, 332-336.

Tallal, P. (1985). Auditory temporal perception, phonics and reading disabilities in children. *Brain and Language, 9*, 182-198.

Vellutino, F.R. (1979a). The validity of perceptual deficit explanations of reading disability: a reply to Fletcher and Satz. *Journal of Learning Disabilities, 12*, 160-167.

Vellutino, F.R. (1979b). *Dyslexia: Theory and Research*, London: M.I.T. Press.

Weisstein, N., Ozoz, G., & Szoc, R. (1975). A comparison and elaboration of two models of metacontrast. *Psychological Review, 2*, 325-342.

Williams, M., & Bologna, N. (1985). Peripheral grouping in good and poor readers. *Perception and Psychophysics, 36*, 367-374.

Williams, M., Brannan, J., & Bologna, N. (1988). Perceptual consequences of a transient subsystem visual deficit in the reading disabled. *Paper presented to the 24th International congress of Psychology, Sydney.*

Williams, M., Brannan, J., & Lartigue, E. (1987). Visual search in good and poor readers. *Clinical Visual Sciences, 1*, 367-371.

Williams, M., & LeCluyse, K. (1990). Perceptual consequences of a temporal processing deficit in reading disabled children. *Journal of the American Optometry Association, 61*, 111-121.

Williams, M., Le Cluyse, K., & Bologna, N. (1990). Masking by light as a measure of visual integration time and persistence in normal and disabled readers. *Clinical Vision Sciences, 5*, 335-343.

Williams, M., Molinet, K., & Le Cluyse, K. (1989). Visual masking as a measure of temporal processing in normal and disabled readers. *Clinical Vision Sciences, 4*, 137-144.

The Role of Eye Movements in Perceptual Processes
E. Chekaluk and K.R. Llewellyn (Editors)

RESPONSE CHARACTERISTICS OF MIDBRAIN NEURONS RELATED TO VERGENCE EYE MOVEMENTS

J.W. Morley

1. Introduction

An essential part of the visual behaviour of animals with binocular single vision is the ability to accurately fixate targets at different depths. When visual fixation is transferred from a far to near target, three oculomotor events take place: the two eyes converge, the dioptric power of the eyes (accommodation) is increased, and the pupils constrict to provide a greater depth of focus. These three motor responses are collectively termed the *near response*, and together allow the target to be focussed on corresponding points of each retina, thereby providing a clear, single image of the target.

Although the neural control of eye movements has been the subject of a great deal of investigation over the last 30 years or so, most work has been directed at elucidating the neural mechanisms underlying conjugate eye movements. Both clinical and experimental evidence suggests that at least at the level of the brainstem the mechanisms controlling vergence eye movements are independent from the circuits involved in the control of conjugate eye movements. For example, the neurological syndrome internuclear ophthalmoplegia, which results from a lesion of the medial longitudinal fasciculus in the brainstem, consists of a paresis or paralysis of adduction of the eyes in conjugate eye movements but not in convergence eye movements (Smith and Cogan, 1959; Cogan, 1970; Gamlin et al, 1989; Leigh and Zee, 1983). Very little investigation however has been directed at the neural mechanisms controlling vergence eye movements. In this review I will outline the response characteristics of neurons in the brainstem that have recently been implicated in the control of vergence eye movements.

2. Simplified Model of Vergence Control

The primary sensory stimulus that drives vergence eye movements is a disparity between the retinal positions of a target's image in the two eyes. Where the retinal disparity is large, vergence responses can be evoked by very brief exposure to visual stimuli, even when the image to each eye is different (Westheimer and Mitchell, 1969). However, for perceptual fusion of a target in depth to occur, the images to each eye must be similar and sustained (Westheimer and Mitchell, 1969; Jones, 1977). The mechanism operating on small retinal disparities to produce a single, fused image is termed *fusional vergence*.

In comparison with conjugate eye movements, particularly saccades, vergence is extremely slow, with a latency to a step change in disparity of approximately 160 ms, and an upper velocity limit of about 10 deg/sec (Rashbass and Westheimer, 1961). Vergence eye movements however are very accurate, with the error between the retinal disparity input and the vergence output - *fixation disparity* - being only a minute or so of arc (Ogle et al, 1949; Rashbass and Westheimer, 1961).

A simplified model of the vergence control pathway is shown in Fig. 1. The

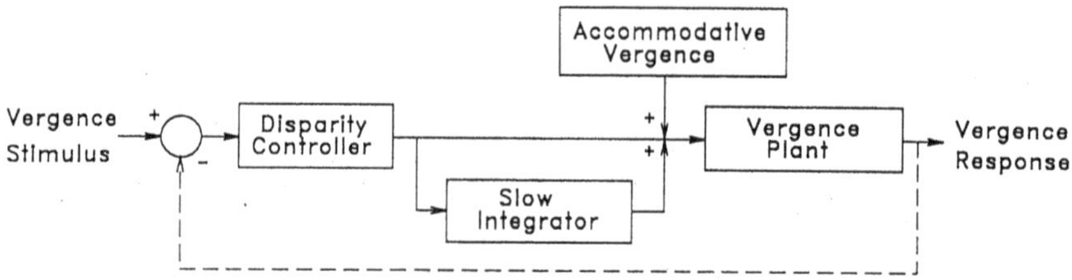

Fig. 1. Simplified model of vergence control The pathway is modelled as a negative feedback controller consisting of two integrators in parallel. Input to the pathway is fixation disparity. The main disparity pathway is modelled as consisting of a leaky integrator (disparity controller) with a decay time constant of approximately 10 - 15 sec. The second integrator (slow integrator) is modelled as having a much longer decay time constant and is responsible for the increase in tonic vergence observed following vergence adaptation. The output of both integrators is to the vergence plant, activation of which results in a vergence movement of the eyes. Input to the main disparity pathway also arises from the accommodation control pathway (not shown) via a blur-driven accommodative-convergence cross-link.

pathway is modelled as a negative feedback controller consisting of two integrators situated in parallel (Westheimer, 1963; Schor, 1979b; Hung and Semmlow, 1980). Input is fixation disparity and output is to the vergence plant, activation of which results in a vergence movement of the eyes. The main disparity pathway is shown as consisting of a disparity controller that is modelled as a leaky integrator with a decay time constant of approximately 10-15 sec., while the second of the two integrators (slow integrator) is modelled as having a much longer decay time constant and is driven by the output of the disparity controller. A third source of input to the vergence control pathway comes from the accommodation control system. It has been known for some time that vergence does not operate independently of accommodation. In monocular viewing, where retinal disparity cues are eliminated, changes in accommodation associated with changing fixation from a far to a near target are accompanied by a change in convergence *(accommodative-convergence)* (Mueller, 1842). [In addition, changes in vergence when viewing binocularly through pinholes, such that blur inputs (the primary sensory stimulus to accommodation) are negligible by producing an infinite depth of focus, are accompanied by a change in accommodation *(vergence-accommodation)* (Fincham and Walton, 1957)]. Figure 1 shows the blur-driven accommodative-convergence cross-link providing input to the vergence plant (the accommodation control pathway is not shown in the figure, however it is analogous to the vergence control pathway; see Schor, 1988). Although the model shown in Fig. 1 is an oversimplification of the vergence control pathway, it serves to indicate the main factors involved in the control of vergence (for a more detailed account of the vergence and accommodation control models see Schor, 1983, 1988).

3. Response Characteristics of Near-Response Neurons

The first systematic study of neurons in the brainstem that are related to the near-response was made by Mays (1984). Mays recorded from midbrain neurons located dorsal and dorsolateral to the oculomotor nucleus, in the awake-behaving monkey, that discharged in relation to the vergence angle of the eyes. By far the most numerous type of cell encountered increased their discharge rate in association with an increase in vergence angle (convergence cells, Fig. 2), but showed no change in activity during conjugate eye movements. A second, less common type of neuron displayed an inverse relationship between firing rate and vergence angle (divergence cell), however the discharge was less highly correlated with vergence angle than was found in convergence cells. Although Mays provided indirect evidence that the midbrain neurons responded to vergence and not accommodation, he did not

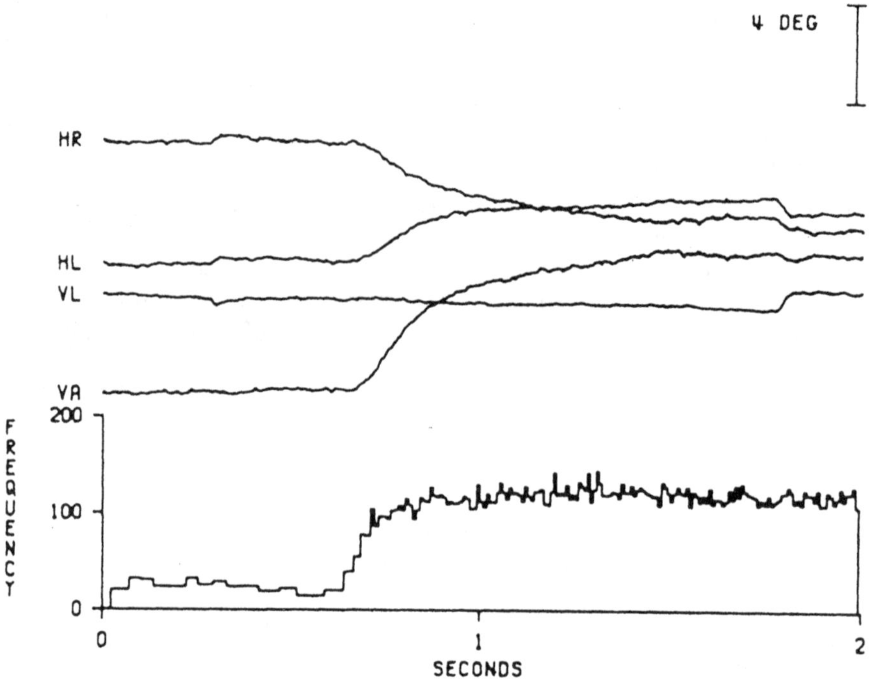

Fig. 2. Response of a midbrain neuron related to the near-response during a vergence movement of the eyes. The monkey was trained to make a 5° symmetrical vergence movement from a far target (72.5 cm from the eyes) to a near target (25 cm from the eyes). The upper part of the figure shows traces of eye position: HR – horizonal position of the right eye; HL – horizontal position of the left eye (rightward movement of the eyes is shown as an upward deflection of the position trace); VL – vertical position of the left eye; VA – vergence angle (difference between HR and HL) with an upward deflection indicating increased convergence. The histogram in the lower part of the figure shows the response of the neuron in spikes/sec (from Mays, 1984).

measure the accommodation and therefore could not rule out the possibility that the neurons were in fact related to the accommodation component on the viewing task.

The findings of Mays were extended by Judge and Cumming (1986). In their study monkeys were trained to visually track targets, presented in a mirror stereoscope (Fig.3), that appeared to move sinusoidally (that is, both vergence and accommodation demand varied sinusoidally) directly toward and away

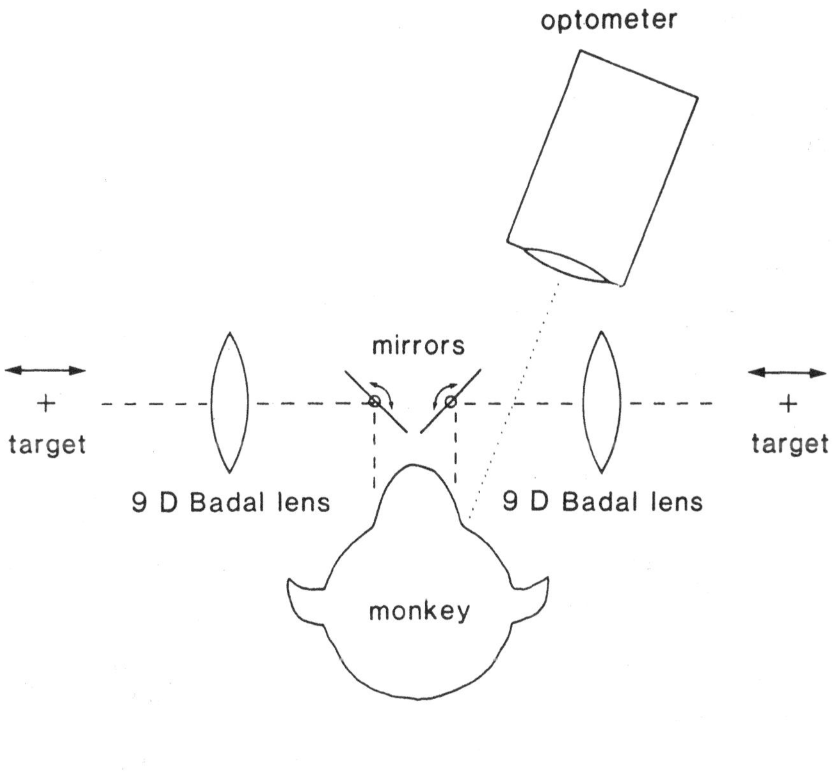

Fig. 3. Schematic representation of a mirror stereoscope. The monkey views targets through Badal lenses via front surfaced mirrors. The Badal viewing system allows the accommodation demand to be altered, by moving the targets towards or away from the animal, without any concomitant change in the angular size of the stimulus. The vergence stimulus is produced by altering the angle of the mirrors. Accommodation is measured with an infra-red optometer.

from the right eye, the left eye or in the midline. The horizontal position of both eyes and the accommodation of the right eye were measured. The firing rate of the midbrain neurons, which they termed *near-response neurons*, was modulated in response to the monkey tracking the target as it appeared to move in depth, but not during dynamic conjugate eye movements. The amplitude of discharge modulation was similar in monocular viewing (blur cue alone), in binocular viewing with accommodation open loop (disparity cue alone) or in normal binocular viewing (both disparity and blur cues present) indicating that the neurons do not respond to sensory cues associated with the near-response (disparity or blur) but rather are related to some motor component of the near-response.

The mirror stereoscope used in the experiments of Judge and Cumming provided independent control over the vergence and accommodation stimuli. This allowed the accommodation and vergence demands associated with viewing a target moving in depth to be partially dissociated *(relative vergence:* Fig. 4). Monkeys were trained to binocularly track a target in which the accommodation demand was half that normally associated with a given vergence demand. This viewing paradigm allowed the relative contributions of vergence demand (retinal disparity) and accommodation demand (blur) to a neurone's response to be determined. To quantify the contributions of each cue, Judge and Cumming used a linear summation model in which the neurone's response, f, was considered to be the linear sum of the vergence and accommodation responses:

$$f = k_a A + k_v V$$

where A and V are the accommodation and vergence responses respectively (see Appendix of Judge and Cumming, 1986 for the method used to derive accommodation and vergence coefficients, k_a and k_v). Near-response neurons with a vergence coefficient (k_v) significantly different from zero, and accommodation coefficient (k_a) not significantly different from zero were classified accommodation had a significant k_a and a non-significant k_v. A number of near-response neurons however had both accommodation and vergence coefficients significantly different from zero, and therefore, appeared to be related to *both* oculomotor responses.

In a recent study Zhang et al (1991b) used a relative vergence paradigm to further investigate the relationship between the response characteristics of near response neurons and accommodation and vergence. To quantify the relationship a linear interaction model similar to that of Judge and Cumming

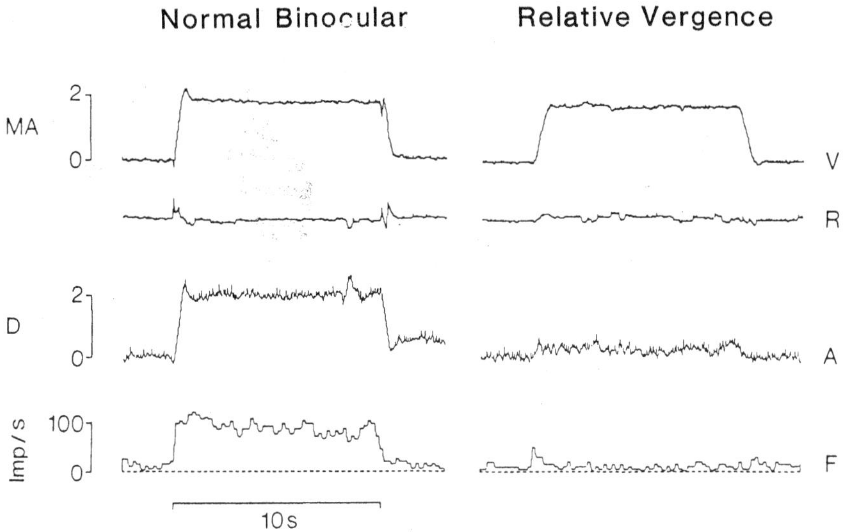

Fig. 4. Response of a midbrain near-response neuron during normal binocular viewing and relative vergence viewing. Eye movement traces are vergence response (V) and horizontal position of the right eye (R) in meter-angles (one meter-angle (MA) is the angle of convergence required to fixate a target at a distance of one meter and is thus commensurate with accommodation in diopters (D)), and accommodation response of the right eye (A). Firing rate (F) of the neuron is in spikes/sec. In A the vergence demand was increased 2MA and the accommodation demand 2D for 10 sec. In B the vergence demand increased 2MA without concomitant change in accommodation demand. The neuron responded with a discharge of up to 100 spikes/sec in normal binocular viewing (A), but displayed virtually no change in response during relative-vergence viewing. This neuron would be judged an accommodation-related neuron.

(1986) was employed. Five categories of near-response neurons were described. One category of neurone was related exclusively to vergence (termed + V cells, 37.5% of sample), as described by Judge and Cumming (1986). The remaining 3 categories of neurons were +V+A neurons (both coefficients significantly different from zero; 34.7% of sample); +V-A neurons (significant positive vergence coefficient and significant negative accommodation coefficient; 19.4% of sample) and +A-V neurons (significant positive accommodation coefficient and significant negative vergence coefficient; 4.2% of sample). These findings are intriguing in that they imply that the *majority of* midbrain near-response neurons are driven by both disparity and blur cues. The positive or negative vergence and accommodation gains of each neuron reflect the disparity or blur input to that neurone relative to the overall vergence or accommodation input.

Zhang et al (1991b) were also able to antidromically activate 17 near-response neurons from the medial rectus division of the oculomotor nucleus. Of these 17 neurons, 4 were classified as +V neurons, 9 as +V+A neurons and 4 as +V-A neurons. These results indicate that the medial rectus motor neurons of the oculomotor nucleus do not receive a *pure* vergence signal from near-response neurons related exclusively to vergence, but rather input from a heterogeneous group of near-response neurons that are driven to a greater or lesser extent by both disparity and blur. The authors suggest however that the positive and negative accommodation gains of the neurons that project to the medial rectus motor neurons of the oculomotor nucleus would tend to cancel, resulting in a *net* vergence signal at the level of the motor neuron.

3.1 Burst and Burst-Tonic Near-Response Neurons

Mays et al (1986) have further described neurons in the midbrain that discharge in relation to the velocity of the vergence movement. One group of neurons, termed *burst neurons*, discharged vigorously at the onset of a vergence movement (Fig. 5). Both convergence and divergence burst neurons were found, with convergence burst neurons far more numerous than divergence burst neurons (ratio of approximately 7:1). The burst neurons discharge in each case was related to the instantaneous vergence velocity, with firing rate proportional to velocity, and led the onset of the vergence movement. Mays et al suggest that the burst neurons in the vergence system play an analogous role to the short-lead burst neurons (previously termed medium-lead burst neurons) in the saccadic system (Fuchs et al, 1985), combining with the vergence integrator in the control pathway (disparity controller in Fig. 1), thereby improving the frequency response characteristics of the system (see Robinson, 1981).

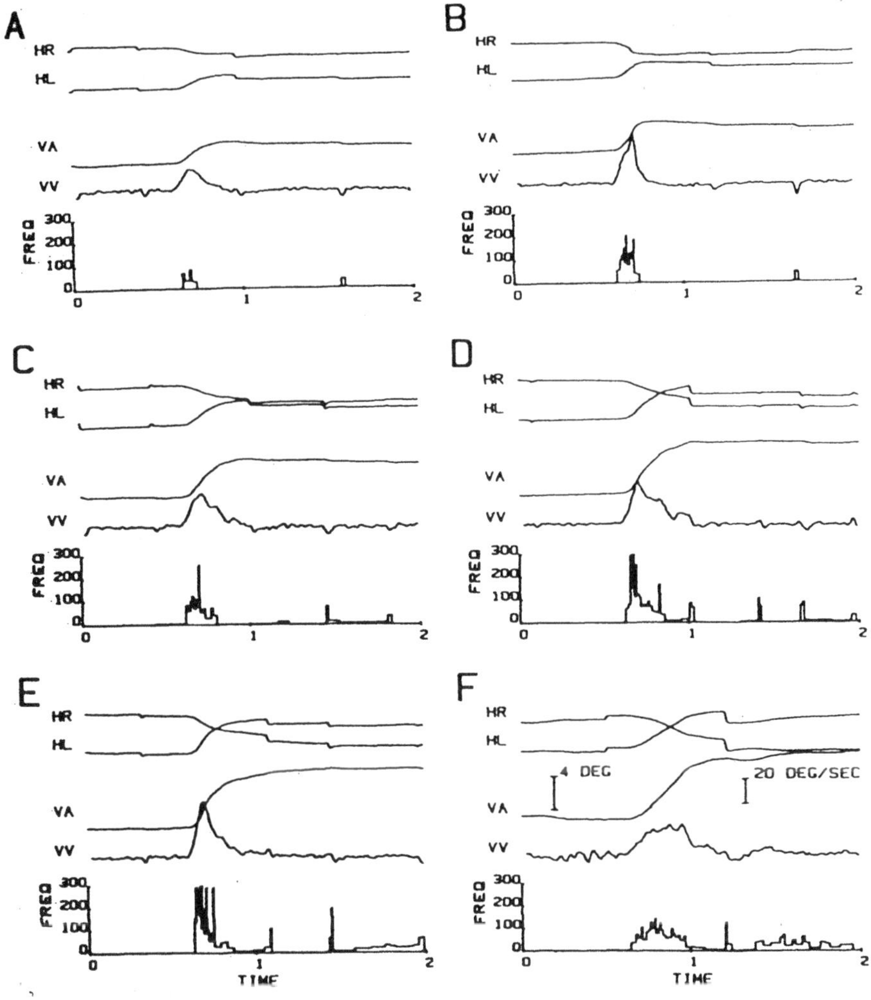

Fig. 5. Relationship between the discharge of a convergence burst neuron and vergence velocity (VV). Eye movement traces HR, HL and VA are as in Fig 2. The amplitude of the symmetrical vergence movement shown in each panel was: 2°in A; 3° in B; 4°in C; 6°in D; 7°in E and 8°in F. The response of the neuron in spikes/sec is shown below each panel. Calibration bars for amplitude and vergence velocity are shown in panel F (from Mays et al., 1986).

A second group of neurons, termed *burst-tonic neurons*, were described in which the discharge of the neurons consisted of a burst of activity in association with the onset of the vergence movement and a tonic discharge once the vergence angle was greater than approximately 4° - 6° (Fig. 6). The presence of burst-tonic neurons in the midbrain has implications in regard to the processing of

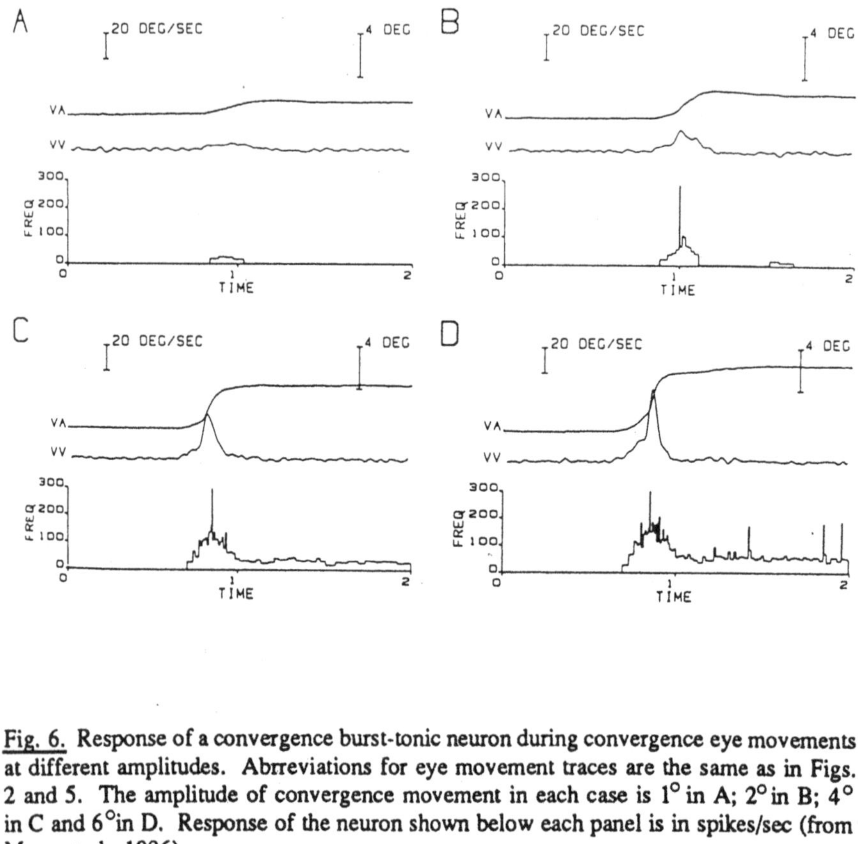

Fig. 6. Response of a convergence burst-tonic neuron during convergence eye movements at different amplitudes. Abbreviations for eye movement traces are the same as in Figs. 2 and 5. The amplitude of convergence movement in each case is 1° in A; 2° in B; 4° in C and 6° in D. Response of the neuron shown below each panel is in spikes/sec (from Mays et al., 1986).

the vergence signal. Gamlin and Mays (1986) have reported that **medial** rectus motor neurons in the oculomotor nucleus show both a vergence velocity and a vergence position signal in their discharge. It would seem that, at least for some cells, the vergence velocity and position signals are **combined prior to** the motor neuron.

3.2 Role of Near-Response Neurons in Vergence Adaptation

In monocular viewing, the vergence position adopted by the covered eye depends on the accommodative state of the viewing eye (accommodative-vergence drive) and any bias in the vergence control pathway. It was shown by Maddox in 1893 that the degree of bias in the vergence control pathway could be modified by the use of prisms. If a wedge prism is placed base-out in front of one or both eyes during binocular viewing extra convergence is required to view a target at a given distance (Fig. 7). If one eye is immediately covered

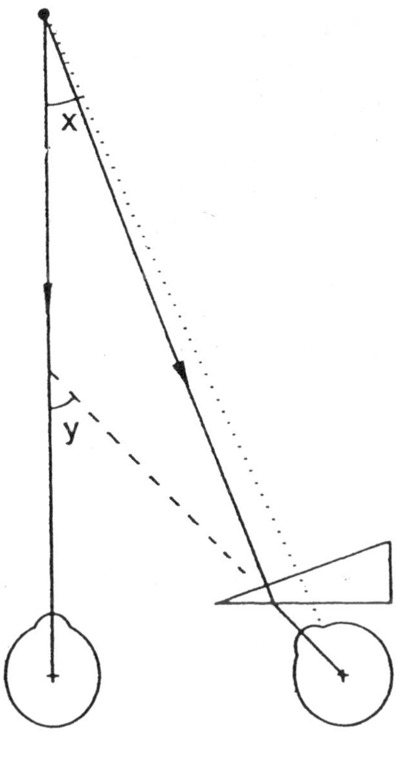

Fig. 7. Representation of the increase in convergence required to view a target when a base-out prism is placed in front of one eye (modified from Judge, 1987).

following the period of prism-viewing the vergence position adopted by the covered eye is increased in a convergent direction. The magnitude, and time course of decay of the after-effect (prism-adaptation) depend on the strength of the prism(s) and the period of prism-viewing (Carter, 1965; Henson and North, 1980; North and Henson, 1981; Miles et al, 1987; Schor, 1979 a & b). Investigation of the phenomenon of prism-adaptation has been carried out predominately in human subjects, however it has also been reported to occur in monkeys following both prism-viewing (Judge, 1987) and manipulation of the vergence demand associated with viewing targets in a stereoscope (Morley et al, 1988).

In relation to the model depicted in Fig. 1, the increase in the resting *(tonic)* level of vergence following prism-viewing is accounted for by the output of the slow integrator. Prolonged viewing of a target where the vergence demand is increased (such as when a base-out prism is placed in front of one eye) results in the slow intergrator being *charged up*. The output of the slow integrator feeds back into the main disparity pathway producing the long lasting increase in tonic vergence observed upon monocular testing. Schor and colleagues (Schor and Kotulak, 1986; Schor et al, 1986) have redefined the dual interaction model of vergence and accommodation control on the basis of their finding that in some subjects vergence bias can be produced in monocular viewing (that is, no disparity cue to vergence), indicating that the slow integrator in the vergence control pathway has both a disparity-driven and a blur-driven input. The redefined model has the accommodative-vergence cross-link positioned before the input to the slow integrator in the vergence control pathway. However, the evidence for accommodation affecting vergence bias is not strong and the position of the cross-link in the pathway remains uncertain.

3.3 Near-Response Neurons and Phoria-Adaptation

The term phoria-adaptation is used here to indicate an increase in the tonic level of vergence. A subject's phoria, measured in monocular viewing, is defined as the deviation of the covered eye from the convergence that would be appropriate to view a target observed by the viewing eye. Adaptation of vergence in the animal studies discussed below was brought about by increasing the vergence demand, relative to the accommodation demand, required to view a target presented in a mirror stereoscope. Phoria-adaptation is therefore a more accurate indication of the experimental manipulation than prism-adaptation, which implies that targets were viewed through prism(s) placed in front of one or both eyes.

If the midbrain near-response neurons play a role in the increase in phoria following adaptation then such increase should be apparent in their response. In particular, if vergence-related neurons are responsible for the observed increase in phoria, their firing rate relative to vergence should increase following adaptation. In contrast, accommodation-related neurons should show no change in their response relative to the accommodation response and a decrease in response relative to vergence (that is, the response function of an accommodation neurone should simply be shifted an amount equal to the change in phoria, resulting in the firing rate at a given vergence angle being decreased in the adapted compared to the unadapted condition).

Tello and Mays (1984a & b) investigated the activity of midbrain near-response neurons in the monkey following phoria-adaptation. They reported that the response of the neurons relative to vergence angle was consistently lower in the adapted condition than in the unadapted condition. Tello and Mays did not measure accommodation response and could not therefore rule out the possibility that they were recording from accommodation-related neurons that maintained a consistent relationship to the accommodation response.

An extended study into the response characteristics of near-response neurons during adaptation in the vergence system was undertaken by Morley et al (1991). As in the Tello and Mays study, monkeys were trained to binocularly view targets in a stereoscope where the vergence demand was increased relative to the accommodation demand. Prior to adapting the monkey, neurons were classified as related exclusively to vergence, accommodation or both using a relative vergence paradigm and employing the linear summation model of Judge and Cumming (1986). The firing rate of neurons classified as accommodation-related maintained an approximately consistent relationship to the accommodation response (Fig. 8A). Although the majority of vergence-related neurons had an increased firing rate relative to vergence angle following adaptation, only a small proportion maintained a consistent relationship to vergence, that is, the increase in discharge of the neurone was sufficient to account for the total increase in phoria. For most neurons however the increase in discharge at a given vergence angle was not sufficient to account for the observed increase in phoria (Fig 8B). On average only 54% of the increase in phoria following adaptation was reflected in the increased discharge of the vergence-related neurons. This figure reduces to 38% if all near-response neurons except those related exclusively to accommodation are included in the average. This finding implies that the vergence bias apparent after adaptation is mediated at more than one site with only a proportion

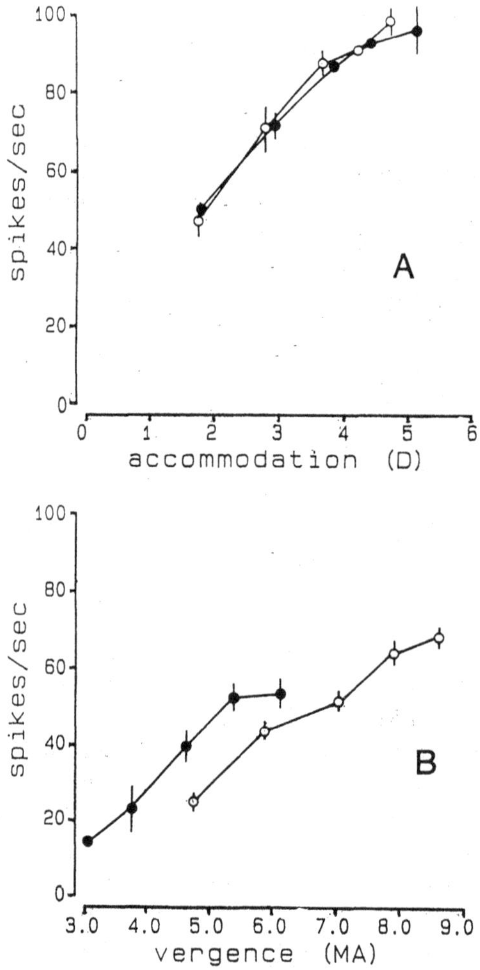

Fig. 8. Response of an accommodation-related neuron (A) and a vergence-related neuron (B) before (filled circles) and after (open circles) phoria-adaptation. Firing rate (spikes/sec) of each neuron is plotted against accomodation in diopters (D) in A and against vergence in meter-angles (MA) in B. The phoria-adaptation had no effect on the response of the accommodation-related neuron, as would be expected if the adaptation was brought about by changes in activity of neurons in the vergence pathway. The firing rate of the vergence neuron in B did increase following adaptation, but not sufficiently to account for the extra vergence. If vergence-related neurons played no part in the observed adaptation, the rate function relating the neurons response to vergence should simply be shifted along the X-axis an amount equal to the increase in phoria produced by the adaptation (from Judge et al., 1991).

of the increase apparent at the level of the midbrain near-response neurons. To achieve the levels of phoria observed following adaptation the proportion of the increased vergence signal in these neurons must be combined with an additional signal, reflecting the remainder of the increase, at a more distal site. The most plausible site is at the level of the motor neurons in the medial rectus division of the oculomotor nucleus (see below).

3.4 Anatomical Considerations

The experiments outlined above have each described the location of the majority of near-response neurons to be within 1-2 mm dorsal and dorsolateral to the oculomotor nucleus. In addition, both Judge and Cumming (1986) and Mays et al (1986) have reported a further group of **midbrain near-response neurons that are located 4-5 mm dorsal to the oculomotor nucleus and up to 3 mm lateral. The location of near-response neurons has been determined using a combination** of identifiable structures, such as the oculomotor nucleus. No anatomical studies have been published in which these or other midbrain neurons possibly involved in the near-response have been observed. It would appear from the data that are available to date that there is in fact no *specific premotor nucleus* for vergence in the midbrain, a function attributed for many years to the central nucleus of Perlia (see Warwick, 1955 for discussion).

3.5 Relationship between midbrain near-response neurons and motor neurons in the oculomotor nucleus

The location and response properties of near-response neurons (Mays, 1984; Judge and Cumming, 1986; Mays et al, 1986) coupled with the latencies to vergence eye movements following microstimulation in the midbrain areas where near response neurons are located (Judge and Cumming, 1986) provide indirect evidence that these neurons are premotor and monosynaptically project to the medial rectus motor neurons of the oculomotor nucleus. Nakao et al (1986) were able to antidromically activate midbrain neurons that project to the medial rectus division of the oculomotor nucleus in cats. However the cats were anaesthetised and paralyzed, making it impossible to determine if the antidromically activated midbrain neurons were in fact near-response neurons.

A recent study by Zhang et al (1991a) provides strong evidence that near-response neurons project monosynaptically to medial rectus motor neurons in the oculomotor nucleus, and that this projection is predominantly, if not completely, ipsilateral. They were able to antidromically activate 28 identified near-response neurons in the awake-behaving monkey from a stimulating electrode placed in the medial rectus division of the ipsilateral oculomotor nucleus. Stimulation of the contralateral oculomotor nucleus did not activate the near-response neurons.

On the basis of anatomical studies, Buttner-Ennever and Akert (1981) suggested that there may in fact be a subgroup of medial rectus motor neurons that are **predominantly** involved in the control of vergence eye movements. Buttner-Ennever and Akert injected retrograde tracer into the orbital (outer) layer of the medial rectus muscle and were able to show labelled motor neurons that formed a subgroup (subgroup C) in the medial rectus divisions of the oculomotor nucleus. The orbital muscle layer contains slow-tonic fibres that are suggested to be suited to carrying out the slow graded types of movement associated with vergence. However it is unlikely that the C subgroup is involved only in vergence eye movements. Porter et al (1983) point out that the C subgroup also receives input from abducens **internuclear** neurons, carrying a signal related to conjugate eye position, and that the orbital layer of the medial rectus muscle contains large, fast-twitch type fibres, suited to carrying out rapid eye movements, along with the smaller slow-tonic fibres.

Physiological evidence supports the idea that motor neurons are differentially sensitive to both vergence and conjugate eye movements. Mays and Porter (1984) investigated the response characteristics of medial rectus motor neurons during vergence and conjugate eye movements and found that the motor neurons were active in each type of eye movement. However, the vergence gain and conjugate gain differed for individual neurons. Motor neurons with a higher vergence gain than conjugate gain would presumably play a greater role in vergence eye movements, and vice versa.

References

Buttner-Ennever, J.A., & Akert, K. (1981). Medial rectus subgroups of the oculomotor nucleus and their abducens internuclear input in the monkey. *Journal of Comparative Neurology*, 197, 17-27.

Carter, D.B. (1965). Fixation disparity and heterophoria following prolonged wearing of prisms. *American Journal of Optometry*, 42, 141-152.

Cogan, D.G. (1970). Internuclear ophthalmoplegia, typical and atypical. *Archives of Ophthalmology*, 84, 583-589.

Fincham, E.F., & Walton, J. (1957). The reciprocal actions of accommodation and vergence. *Journal of Physiology (London)*, 137-488-508.

Gamlin, P.D.R. and Mays, L.E. (1986). Medial rectus motoneurons carry a vergence volocity signal in addition to a vergence position signal. *Society for Neuroscience Abstracts*, 12, 460.

Gamlin, P.D.R., Gnadt, J.W., & Mays, L.E. (1989). Lidocaine-induced unilateral internuclear ophthalmoplegia: Effects on convergence and conjugate eye movements. *Journal of Neurophysiology*, 62, 82-95.

Henson, D.B., & North, R. (1980). Adaptation to prism-induced heterophoria. *American Journal of Optometry*, 57, 129-137.

Hung, G.K., & Semmlow, J.L. (1980). Static behavior of accommodation and vergence; computer simulation of an interactive dual-feedback system. *IEEE Transcripts in Biomedicine and Engineering*, BME-27, 439-447.

Jones, R. (1977). Anomalies of disparity detection in the human visual system. *Journal of Physiology (London)*, 264, 621-640.

Judge, S.J. (1987). Optically-induced changes in tonic vergence and AC/A ratio in normal monkeys and monkeys with lesions of the flocculus and ventral paraflocculus. *Experimental Brain Research*, 66, 1-9.

Judge, S.J., & Cumming, B.G. (1986). Neurons in the monkey midbrain with activity related to vergence eye movement and accommodation. *Journal of Neurophysiology*, 55, 915-930.

Judge, S.J., Cumming, B.G., Morley, J.W. & Lindsey, J.W. (1991). The role of midbrain neurones in controlling accommodation and convergence of the eyes. In M. Yoshikawa, M. Uono, H. Tanabe and S. Ishikawa (Eds), *New trends in autonomic nervous system research: basic and clinical integration*. Elsevier, Amsterdam, pp.132-135.

Leigh, R.J., & Zee, D.S. (1983). *The neurology of eye movements*. F.A. Davis Company Philadelphia.

Maddox, E.E. (1893). *The clinical use of prisms*. 2nd Ed. John Wright and Sons, Bristol.

Mays, L.E. (1984). Neural control of vergence eye movements: convergence and divergence neurons in midbrain. *Journal of Neurophysiology*, 51, 1091-1108.

Mays, L.E., & Porter, J.D. (1984). Neural control of vergence eye movements: activity of abducens and oculomotor neurons. *Journal of Neurophysiology*, 52, 743-761.

Mays, L.E., Gamlin, P.D.R., & Tello, C.A. (1986). Neural control of vergence eye movements: neurons encoding vergence velocity. *Journal of Neurophysiology*, 56, 1007-1021.

Morley, J.W., Judge, S.J., & Lindsey, J.W. (1991). Role of monkey midbrain near-response neurons in phoria adaptation. Submitted to *Journal of Neurophysiology*.

Morley, J.W., Lindsey, J.W., & Judge, S.J. (1988). Prism-adaptation in a strabismic monkey. *Clinical Vision Science*, 3, 1-8.

Mueller, J. (1843). *Elements of physiology* (translated by W. Baly) London: Taylor and Walton, Vol.2.

Nakao, S., Shiraishi, Y., & Miraya, T. (1986). Direct projection of cat midbrain tegmentum neurons to the medial rectus subdivision of the oculomotor complex. *Neuroscience Letters*, 64, 123-128.

North, R., & Henson, D.B. (1981). Adaptation to prism-induced heterophoria in subjects with abnormal binocular vision or asthenopia. *American Journal of Optometry*, 58, 746-752.

Ogle, K.N., Mussey, F. & Prangen, A. (1949). Fixation disparity and the fusional processes in binocular single vision. *American Journal of Ophthalmology*, 32, 1069-1087.

Porter, J.D., Guthrie, B.L., & Sparkes, D.L. (1983). Innervation of monkey extraocular muscles: localization of sensory and motor neurons by retrograde transport of horseradish peroxidase. *Journal of Comparitive Neurology*, 208-219.

Rashbass, C., & Westheimer, G. (1961). Disjunctive eye movements. *Journal of Physiology (London)*, 159, 339-360.

Robinson, D.A. (1981). Control of eye movements. In: *Handbook of Physiology. The nervous system*. Ed. by J.M. Brookhart and V.B. Mountcastle. Bethesda, MD. American Physiology Society, Sect. 1, Vol.1, pp.1275-1320.

Schor, C.M. (1979a). The influence of rapid prism adaptation upon fixation disparity. *Vision Research*, 19, 757-765.

Schor, C.M. (1979b). The relationship between fusional vergence eye movements and fixation disparity. *Vision Research*, 19, 1359-1367.

Schor, C.M. (1983). Fixation disparity and vergence adaptation. In: *Vergence eye movements: Basic and clinical aspects*. Eds. C.M. Schor and K. Ciuffreda. Butterworths, Boston, pp.465-516.

Schor, C.M. (1988). Phasic and tonic organization of accommodation and vergence. In: *Strabismus and amblyopia*. Eds. G. Lennerstrand, G.K. von Noorden and E.C. Campos. Plenum Press, N.Y., pp.111-120.

Schor, C.M., & Kotulak, J.C. (1986). Dynamic interactions between accommodation and convergence are velocity sensitive. *Vision Research*, 26, 927-942.

Schor, C.M., Kotulak, J.C., & Tsuetaki, T. (1986). Adaptation of tonic accommodation reduces accommodative lag and is masked in darkness. *Investigative Ophthalmology and Visual Science*, 27, 820-827.

Smith, J.L., & Cogan, D.G. (1959). Internuclear ophthalmoplegia. A review of 58 cases. *Archives of Ophthalmology*, 61, 687-694.

Tello, C.A., & Mays, L.E. (1984). Activity of the convergence neurons during prism adaptation. *Proceedings of the International Society for Eye Research*, 3, 60.

Tello, C.A., & Mays, L.E. (1984). Activity of mesencephalic convergence cells during vergence adaption. *Society for Neuroscience Abstracts*. 10, 988.
Westheimer, G. (1963). Amphetamine, barbiturates and accommodative convergence. *Archives of Ophthalmology*, 70, 830-836.
Westheimer, G., & Mitchell, A.M. (1969). The sensory stimulus for disjunctive eye movements. *Vision Research*, 9, 749-755.
Zhang, Y., Gamlin, P.D.R., & Mays, L.E. (1991a). Antidromic identification of midbrain near response cells projecting to the oculomotor nucleus. *Experimental Brain Research*, 84, 525-528.
Zhang, Y., May, L.E., & Gamlin, P.D.R. (1991b). Characteristics of near response cells projecting to the oculomotor nucleus. Submitted to *Journal of Neurophysiology*.

The Role of Eye Movements in Perceptual Processes
E. Chekaluk and K.R. Llewellyn (Editors)

EYE MOVEMENTS: DEVELOPMENTAL ASPECTS

Sandra L. Shea

"What one knows is, in youth, of little moment; they know enough who know how to learn."

Henry Adams

1. Introduction

Parents of infants know that even neonates can move their eyes. They also know that these eye movements change and improve rapidly over the first few months of life. Not content with these casual observations, vision researchers have greatly increased their systematic investigations in the last few decades into the different types of eye movements that human infants show at diverse ages, both pre- and post-natally. They have attempted to quantify and assess infants' capabilities to scan, localize and follow targets both moving and stationary in the world around them. As pointed out in earlier work (Teller & Movshon, 1986), one of the great, final truisms of developmental psychology is that babies get better as they get older. Undaunted by that indisputable point, this chapter will explore the differential trends in the development of the various types of eye movements infants produce as they get older and "better".

The recent explosion in research on all aspects of perceptual development has been based on, and augmented by, the rapid development, improvement and availability of methodology and hardware. In turn this has allowed researchers to implement questions and gain insights which are increasingly more sophisticated and more practical (some recent reviews of the field: Aslin, 1987b; Aslin & Smith, 1988; Banks & Salapatek, 1983; Boothe, Dobson & Teller, 1985; Salapatek & Banks, 1978; Salapatek & Cohen, 1987a,b; Teller & Movshon, 1986). This work has delineated the changing capacities of human infants in great detail, but questions remain about the developmental patterns of motor and sensory abilities and their limitations.

Research on the adult oculomotor system frequently concentrates on the precise responses made in the first 100 msec after target displacement, the

Chapter preparation supported by NSF grant BSN-8919865. The author extends appreciation to J. Kidd and S. Mendez for their bibliographic assistance, and to R. Aslin and J. Dannemiller for helpful comments and insights.

nuances of velocity changes over the course of a single saccade, or the small corrections made at the end of a saccade. Because they are dealing with a mature and well-researched system, these researchers can now routinely make assumptions about the basic system, i.e., the ability to move the two eyes concomitantly and accurately (for recent review see Hallett, 1986). Researchers working with the deficiencies associated with traumas, strokes, motor imbalances, or neuromuscular diseases, however, concentrate in detail on how these abnormalities manifest themselves in a mature system (Collewijn et al., 1988; Thaker et al., 1989; Gay et al., 1974; Leigh & Zee, 1984; Leigh, 1989; Michaels, 1985). Researchers working with a normally developing infant's visual system are confronted with some aspects of the system which are strikingly mature even at a very young age, and others which are immature but which become more proficient in a short period of time (for reviews see Aslin, 1987a; Boothe et al., 1985; Shupert & Fuchs, 1988).

Researchers are frequently confounded by the puzzle of how the oculomotor system works as well as it does at such young ages, given that so many system components change, i.e., the extraocular muscles, retina, cortex, geniculate, colliculus, interpupillary distance. Nor is it known how these constantly changing components interact with processing the incoming stimuli and outgoing motor control as well as with each other. Researchers also wonder why the system is inaccurate in the particular patterns it shows before maturity and what makes the system fail in such striking clinical abnormalities as strabismus, either with or without amblyopia.

This chapter concerns itself with the inaccuracies of the developing oculomotor system which could stem from anatomical, physiological, attentional, sensory or motor deficits or a combination of limitations in several or all components of the visual system. One practical aspect to exploring these limitations is the necessary restriction of investigations of behavioral visual performance in human infants. It is not possible to perform routine basic physiological and anatomical work on human infants which would provide information about the structures and neuronal capabilities underlying performance. The exceptions (i.e., Abramov, et al., 1982; Hickey, 1977; Hitchcock & Hickey, 1980; Isenberg, 1986; Yuodelis & Hendrickson, 1986) have contributed greatly to our understanding of the immature system, but compared to our knowledge of the mature human system, there are many unknowns, and compared to our knowledge of the mature system of nonhuman primates, we know very little indeed. In fact, we know little about oculomotor performance in nonhuman infant primates because finding a workable testing situation is so difficult. This practical restriction to behavioural work limits the conclusions researchers can draw about underlying structures and activity, but provides considerable information about the developing capabilities of the human infant.

2. Developmental Problems of the Oculomotor System, Physiology and Anatomy

The anatomy and physiology of the mature visual system and the oculomotor subsystem are enormously complex. The saccadic system alone has been the subject of considerable recent research (Buttner-Ennever, 1988; Wurtz & Goldberg, 1989). Summaries of the burgeoning literature presented in advanced neuroanatomy and neurophysiology texts (i.e., Nieuwenhuys, Voogd & Huijzen, 1988) serve to both astound and depress - the former because so much has been explored and worked out in detail, and the latter because so much is not yet understood: "(we) still lack a coherent picture of the central neural mechanisms for the control of eye movements" (Nieuwenhuys et al., 1988, p. 185). Compared to what is known about the mature system, the anatomical and physiological immaturities of the human infants' system have been little studied (understandable given the requirements of a chronic preparation) and, thus, the task of understanding how and what infants see, and at what age, becomes daunting. Some basic descriptions of what we know about the developing oculomotor system help define the context of the problem.

A. The Problem of Generating Eye Movements

When the problem of developmental eye movements is first approached, it is easy enough to conceptualize it as "infant must follow or maintain target". An infant fixating a point straight ahead or a stationary target elsewhere shows a balanced state of all the extraocular muscle pairs: medial and lateral rectus, superior and inferior rectus and superior and inferior obliques (Fig. 1A and Table 1). Within each pair of muscles the axis of rotation is essentially the same, and therefore their action is antagonistic. In the schematic of fixation shown in Fig. 1A, somewhat ironically, the first condition of activity of the extraocular muscles is no motion[1]. The function of the extraocular muscles in this condition is not to maintain fixation by

[1] "no motion" here is a relative term. The eye is not really still since microsaccades and constant small movements of the eyes occur even in adults. However, the situation as described here means no voluntary or elicited motion in response to a target.

moving the eyes, but rather to maintain fixation by *not* moving the eyes. The muscles here are not inactive, but rather they are maintaining whatever position of bifoveal fixation was initially present.

How does a stationary infant efficiently "follow a target visually", where "follow" means "maintain bifoveal fixation?" When a target moves across the visual field from left to right and the infant's head is vertical (either because the infant is holding its head vertical or their head is being held vertical), a complex set of behaviors is initiated. To follow the target, the infant must relax the OS[2] lateral rectus (LR), and contract the OS medial rectus (MR) (Fig. 1B). The infant must provide "non-interference" from the superior oblique (SO), inferior oblique (IO), superior rectus (SR) and inferior rectus (IR) muscles. That is, these four muscles must maintain their action to keep the eye from drifting up or down, or changing torsion angle, but the OS obliques must increase their action somewhat to assist the globe in moving right (adduction) and to maintain torsion angle. Similarly, the OD LR must contract and the OD MR must relax. The OD SR and IR must maintain their action and "non-interference", and the obliques must decrease pressure to allow the globe to move right (abduction). All muscles of *both* eyes must perform in unison if the target is to remain foveated.

If the target is following a vertical path, all 12 extraocular muscles are still involved, but the ratio of activity is reversed (Fig. 1C). Instead of the two sets of lateral recti relaxing and contracting and the superior and inferior recti and obliques providing "non-interference", for the eye to move vertically the lateral recti must provide "non-interference" support whether the eye is moving up or down. If the eye is moving up, the SR and SO muscles contract to pull the eye up, and the IR and IO relax. Inversely, the IR and IO contract to pull the eye down and the SR and SO relax.

The distance of the target adds yet another layer of complexity. Following a target left and right or up and down in the frontoparallel plane requires cooperation of the 12 extraocular muscles. In these conjugate movements, the eyes are moving together and a similar signal is being sent, and acted on, by the symmetrical muscle pairs involved. If the OS's LR is relaxed, then its MR is contracted. In such a case the OD's LR is contracted and its MR is relaxed. However, if the target is moving *toward* the infant or *away* from the

2 OS is the abbreviation for the left eye (Latin = Oculus Sinister). OD, then, is the right eye (Latin = Oculus Dexter). OU is both eyes (Latin = Oculus Uterque).

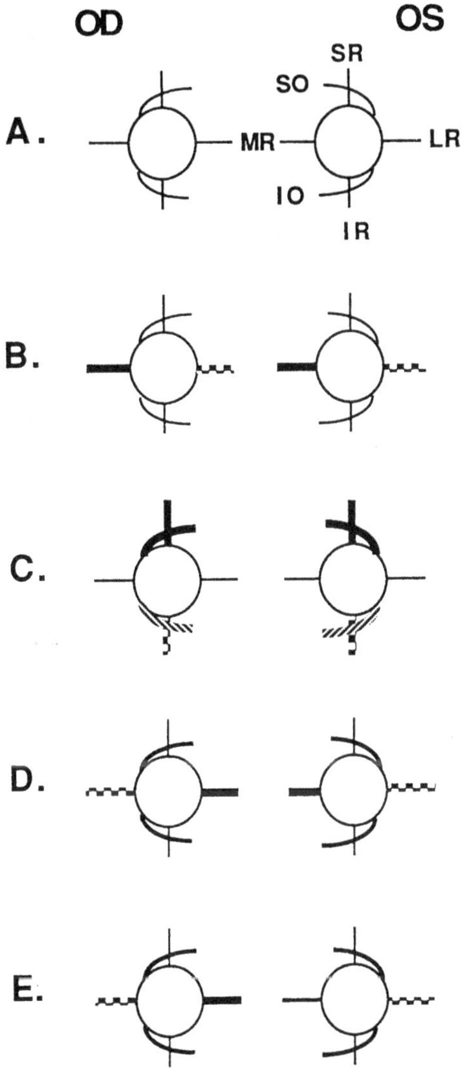

Fig. 1. Schematic of oculomotor muscle demand and effort. (A) At rest, or when fixating a centrally located non-moving target. All 12 muscles are stationary. SR = superior rectus. SO = superior oblique. MR = medial rectus. IO = inferior oblique. IR = inferior rectus. LR = lateral rectus. (B) Response to target moving right. (C) Response to target moving up. (D) Response to target moving in on midline. (E) Response to target moving in off-midline, on left side. Schematics assume infant's head is vertical. Thin solid lines indicate balance or "non-interference". Dotted lines indicate "release" action. Thick solid lines indicate "tightened"/"pull" action.

Table 1
Anatomy and Primary Functions of Extraocular Muscles

Muscle	Length (mm)	Direction of Pull (deg)	Primary Function	Secondary Function	Tertiary Function
Medial rectus	39	90	adduction (turn eye inward)	- - - -	- - - -
Lateral rectus	40	90	abduction (turn eye outward)	- - - -	- - - -
Superior rectus	40	23	elevation	intorsion	adduction
Inferior rectus	40	23	depression	extorsion	adduction
Superior oblique	60	51	intorsion	depression	abduction
Inferior oblique	37	51	extorsion	elevation	abduction

From Michaels, 1985 and Leigh & Zee, 1984

infant, the pattern changes. Consider a target moving toward the infant on an axis aligned with the infant's nose (Fig. 1D). The infant may be able to change vergence enough to follow the target. But following a target moving in depth requires a different set of commands. A target moving towards the infant requires the LR of both eyes to relax while the MR of both eyes must contract. Even in this situation the extraocular muscles' reactions to the moving target are symmetrical. Now consider the situation where the target is moving towards the infant, but is almost directly in front of the left eye (Fig. 1E). Here, the infant must respond *asymmetrically* to maintain bifoveal fixation successfully. The lateral recti of both eyes must still relax and the medial recti of both eyes must still contract, but if the target is to remain foveated, OS needs to rotate in much less than OD.

When the target is not moving and the infant *is* moving, consistent fixation is maintained by visual and non-visual signals to the extraocular muscles mediated by the vestibular system which will be addressed later. An essential part of this system is that there are 3 semicircular canals in each vestibular labyrinth: anterior, posterior and horizontal. The pairs of extraocular muscles are laid out in angles which are most nearly parallel to the ipsilateral semicircular canal which provides the primary excitatory input to that muscle pair (Nieuwenhuys, et al., 1988). In humans, as shown in Table 1, the angle of pull from the midline is 23° for the superior/inferior recti, and 51° for the obliques. The corresponding angles of the anterior and posterior semicircular canals are 41° and 56° respectively (Simpson & Graf, 1981). While the location of the eyes and structure of the extraocular muscles and canals changes across front and lateral-eyed species, this relation remains essentially constant (Hepp, et al., 1989; Simpson & Graf, 1981) and forms the basis for the communication involved in the vestibular ocular reflex (VOR).

The description above is still too simplified to convey the problem adequately. If an infant is going to move his or her eyes to follow or localize a target, the following things must happen: the image of the target must be detectable by the retina. The retina's image must first be transmitted to a "command center", be it collicular, cortical or some other locale. The "command center" must send the correct signal to the oculomotor control center. The oculomotor control center must send the correct message to the extraocular muscles in both eyes. The extraocular muscles must be able to receive this message. The extraocular muscles must be able to act on this message. Somewhere in this process, the target's visual direction must be decoded and translated into the oculomotor task necessary to fixate it, and this is no small task for a mature system (Sparks, 1989), let alone an immature system (Aslin, 1989). Anywhere in those paths - be it ascending information about the target location, or descending information about muscle control - *anywhere* that a signal is delayed or misconstrued, the task of "infant must follow target" becomes, at best, inaccurate, and, at worst, unexecutable. Finally, just finding the target is not sufficient because once "found", it must remain foveated. Thus, any eye movement, or series of eye movements, which is begun, must also be stopped, and stopped accurately. In a saccadic eye movement, for example, the eye is not stopped by the activation of the opposite muscle of the pair. Rather, the eye slows by deceleration caused by the mass of the eye itself and by viscosity. Once stopped, the eye is held in place by the muscles until the next eye movement.

This is an enormously complex set of tasks with which to saddle an immature system, and it is not at all surprising that young infants respond to these stimulus situations in ways that are different from adults. The amazing thing, in fact, is that they respond so similarly at such an early age.

B. The Problem of Developmental Anatomy and Physiology

While adults have to co-ordinate the same muscles to respond to the same stimuli that the infants do, they have the double benefit of thousands of previous executions of similar tasks *and* a visual system which is basically unchanging. For the human infant, however, while all the major anatomical units of the visual system are in place at birth, neither the retinal-geniculate-cortical pathway nor the extrageniculate visual pathway (retina to superior colliculus to pulvinar to primary and secondary visual cortex) are in their final anatomical configuration, let alone functioning at an adult level. Consider some of the immaturities which are present in the infant's visual system. The eye is fairly well developed at birth, but growth and refinement still takes place postnatally (Green, Powers & Banks, 1980; Hansen & Fulton, 1989; Isenberg, 1989; McBrien & Barnes, 1984). The eyeball is approximately 70% of its adult size at birth, and 95% at 3 years, with adult size being attained in early adolescence (Larsen, 1971a, b, c, d; Telang & Dweck, 1989). The cornea is approximately 80% of its adult size at birth and 95% at 1 year, and errors in the young cornea contribute most of the astigmatism seen in young infants (Howland, 1982; Howland et al., 1978; Fulton et al., 1980; Mohindra et al., 1978). The human retina is still undergoing significant development at birth. Cell growth and cell death have already occurred in the ganglion cell layer (Provis, 1987), but retinaletinal vascularization shifts significantly as the hyaloid vessels decrease and mature retinal vascularization increases (Cogan & Kawabara, 1986; Torczynski,

1989) and this changeover is still in progress at birth. Ophthalmoscopic exam shows the macular pigmentation, annular reflex of the macula and the concavity of the foveal pit appear to be generally complete before birth in full-term infants, but the foveal light reflex may not be present until 40 weeks gestation, or even a few weeks later (Isenberg, 1986). At the cellular level, however, while the peripheral retina's structure is in place at birth, photoreceptors and ganglion cells take months to migrate into their final position in the mature fovea (Abramov et al., 1982; Aslin, 1985, 1987c, 1989; Hendrickson & Yuodelis, 1984; Isenberg, 1986; Yuodelis & Hendrickson, 1986) and the retina's neurotransmitters are also immature at birth (Hollyfield, Frederick & Rayborn, 1983). The optic nerve's length and diameter increase during the first 6 months postnatally (Torczynski, 1989) and the myelination of the optic nerve is not completed until after 1 year (7 months to 2 years) (Telang & Dweck, 1989). The iris pigmentation is not stable until approximately 6 months (Telang & Dweck, 1989), though some Caucasian infants show significant changes in eye color in their second year.

The lens of an infant's eye is well-formed and is actually quite powerful. It is so flexible that infants can accommodate (focus) on targets only 6 to 7 cm away, while the lenses of adults in their 30's have already lost some flexibility and may only be able to focus in to distances of 12 or 15 cm. The larger the range of accommodation, the wider the range of near and far targets that can be brought into focus. Infants have a range of accommodation of 15 diopters[3] while 30 year olds may have 7 (Weale, 1990). The ability of infants to change their accommodation accurately in response to targets at different distances has been measured using a variety of techniques designed specifically for infants (Haynes, White & Held, 1965; Braddick et al., 1979; Banks, 1980a, b; Brookman, 1983; Sokol, Moskowitz & Paul, 1983; Howland & Sayles, 1984; Howland, Dobson & Sayles, 1987; Atkinson et al., 1988; Aslin, Shea & Metz, 1990; Riddell et al., 1991; Shea, 1991). Neonates are not very accurate when they attempt to focus targets. At a few months of age they shift accommodation in the right direction when a new target is introduced, but not necessarily by the right magnitude. By 4 to 6 months they are both rapid and accurate in their response to stationary or moving targets.

The extraocular muscles are formed early in ontogeny with the rectus muscles forming and attaching before the obliques (Evinger, 1988;

[3] diopter - a measure of the focal length of a lens. The power of a lens in diopters = 1/(focal length in meters). The more the lens of the eye can bend or refract the light, the higher the power.

Torczynski, 1989). Of the 6 extraocular muscles, 5 originate in the back of the eye's orbit, at its apex, from a common ligament, the annulus of Zinn. The sixth muscle, the inferior oblique, originates anteriorly, at the nasal rim (Hepp et al., 1989; Michaels, 1985; Spencer & Porter, 1988). The four recti are attached anterior to the eye's equator and exert their action by pulling backward. The two oblique muscles are attached posterior to the eye's equator and pull forward.

The structure of the orbit changes postnatally as well. Not only does its overall size increase, but its vertical diameter increases dramatically as compared to its horizontal diameter between the ages of 6 and 12 months (Isenberg, 1988). The effective visual field in very young infants is $\pm 28^{\circ}$ left and right, and 11° up, and 16° down from center. It also increases postnatally, with rapid increases after 2 months. At approximately 1 year the field is essentially adult-like, although the lower and horizontal axes still increase somewhat after this age (Mohn & Van Hof-van Duin, 1986).

The lateral geniculate nucleus (LGN) of the thalamus (Schlag-Rey & Schlag, 1989) is an important waystation of the visual system and is also not complete at birth (Hickey, 1977; Hickey & Peduzzi, 1987; Hitchcock & Hickey, 1980). The optic tract projects from the retina to the LGN, which then projects to the primary visual cortex. The LGN first appears at 10 weeks of gestation and its distinctive lamination takes place between weeks 22 and 25 of gestation (Hitchcock & Hickey, 1980). The topographic representation of the visual field, especially of the optic disc which is found in layers 4 and 6, also occurs at this time, but considerable cell growth occurs in the last 10 weeks of gestation. At birth the LGN is well laminated and the projection of the optic disc is well developed (Hickey & Guillery, 1979), but significant cell growth takes place in the first 6 postnatal months, and some continues through the first 2 years (Hickey, 1977).

Another important waystation in the visual system is the superior colliculus (SC). It receives both sensory and motor signals from cortical and subcortical areas (Grantyn, 1988; Kennard, 1986; Nieuwenhuys, et al., 1988; Sparks & Hartwick-Young, 1988), including the frontal eye fields, auditory and motor cortices. The SC maintains a retinotopic map of the visual fields, has a topographical connection with the pulvinar (Robinson & McClurkin, 1988), and receives auditory information from the inferior colliculus. The pars reticulata of the substantia nigra (SNr) is the only basal ganglia area which sends input to the SC, and may be the only inhibitory input the SC receives (Hikosaka, 1989; Hikosaka & Wurtz, 1989) and thus may regulate input from other areas. The cerebral cortex and the thalamus both have reciprocal connections with the basal ganglia (Nieuwenhuys et al., 1988), and the cerebellotectal pathways are involved with collicular control (May et al., 1990). Motor signals about saccades are sent from the SC to the brainstem reticular formation and then on to the extraocular motoneurons (Wurtz & Albano, 1980). Among other things, the SC is involved with saccade averaging (van Opstal & van Gisbergen, 1990).

The cerebellum, which consists of the cerebellar cortex and nuclei, also co-ordinates information from multiple sensory signals. It is heavily involved with motoric processing and temporal computation, possibly involved with gain control (Gilbert & Thach, 1977; Keller, 1988), which means co-ordinating the timing of motor output to match target location, and the modulation of the accuracy of saccades and other motor activities (Blanks, 1988; Keele & Ivry, 1990; Keller, 1988; Llinas & Pellionisz, 1985; Nieuwenhuys et al., 1988). It begins its development later in ontogeny than other parts of the visual pathway, including the brainstem, but finishes development earlier (Dobbing & Sands, 1973), though it is not complete at birth. The cerebellum retains some plasticity during development, and on into adulthood, a characteristic which enables it to respond to the changing size of body parts, proportions and weights as well as with aging and injury (Llinas & Pellionisz, 1985). Along with the brainstem vestibular nuclei, the cerebellum receives input from the labyrinth by way of the eighth cranial nerve and the climbing fiber system. Like the cerebellum, the inferior olive (InfO) of the brainstem receives information from the climbing fiber system and fires in synchronous patterns. The InfO is known to be sensitive to retinal slip information and vestibular stimulation (Llinas & Pellionisz, 1985; Nieuwenhuys et al., 1988).

The vestibular system itself starts to form quite early in human gestation, around 7 - 8 weeks, and at 16 weeks gestation, when myelination of the central nervous system begins, one of the first myelinated points is the vestibular nerve (Bradley & Mistretta, 1975). Its anatomy is complete, and the system functional at birth (Ornitz et al., 1979; Ornitz, 1983), though postnatal changes do occur.

The vast majority of information about the development of the cortex comes from work with non-human primates. The classic work of Conel (1939, 1941, 1947) provides some of the most extensive work on human cortical development. More recently, researchers have found that the visual cortex itself has more than its full complement of cells at birth, and in fact, has lost thousands of cells during the last half of gestation (Leuba & Garey, 1987) and continues to lose more until about 4 months after birth. Synaptic connections, however, continue to increase throughout the first year of postnatal life (Michel & Garey, 1984). The development of the ocular dominance columns appears to be immature at 4 months, but well developed by 6 (Hickey & Peduzzi, 1987).

Many components of the human oculomotor system, then, are immature at birth, and developmental changes take place throughout the first months, and even years, of life. How and when these changing components mature, and how they interact with each other in the generation and control of eye movements continues to be a topic of intense research.

3. Current Methodology

A. Eye Movement Measurement Systems

The two most common methods and procedures currently used to measure eye movements in human infants are the electrooculogram (EOG) and corneal reflection (CR) systems. The elaborate high-tech and high resolution systems such as the magnetic search-coil and the SRI Purkinje image tracker are unsuitable for use with infants. The search coil systems require the placement of contact lenses or the implantation of coils in subjects' eyes. This is, at best, impractical with human infants. The SRI Purkinje tracker requires a stable image to "lock on" to its target, thus requiring a subject to remain motionless for the better part of a minute each time the tracker needs to regain the image. The SRI Purkinje tracker also uses all four Purkinje images, rather than the first image (the corneal reflection) used by other systems (Crane and Steel, 1985). To generate the fourth Purkinje image, the light source must be bright enough to overcome the opacities of the optic media. While the light is safe for use with adult humans and other primates, its effect on the infant's immature retina is unknown. An intriguing procedure using Doppler ultrasound techniques to measure muscle velocities is still in the developmental stage (Canning & Restori, 1989). Unfortunately these restrictions eliminate the two systems with the highest spatial and temporal resolution, but the remaining systems have provided significant qualitative and quantitative data on the development of eye movements.

As with many dependent measures in experimental psychology, each technique of eye movement recording has its disciples and its detractors. What is critical, however, is that *the relative advantages and disadvantages of each technique be weighed within the context of the question being asked.* For example, the free head movement required to record VOR has prohibited the use of CR systems. On the other hand, one cannot elminate the muscle and head movement contamination inherent in EOG. This does not mean that both systems should be abandoned. It does mean that the tool must be matched to the task. This problem is considered in detail below.

Electrooculography - EOG uses surface electrodes placed near the orbit of one or both eyes. Horizontal eye movements can be recorded by locating the electrodes on the external and internal canthi of one eye or on the external canthi of the two eyes. If vertical eye movements are of interest, the electrodes are placed above and below one eye. EOG records the electrical potential between the cornea and the retina. This corneal-retinal potential is altered with respect to the fixed electrodes when the eye rotates in the orbit, thus *measuring eye position relative to head position*. Its signal is generally linear with peripheral targets located +/-20 to 30 deg from a central fixation point (i.e., Aslin & Salapatek, 1975; Young & Sheena, 1975; Shupert & Fuchs, 1988; Finocchio, Preston & Fuchs, 1990), though it may not be linear in all directions for all subjects (Maurer, 1975b). EOG can measure saccades made to a target displaced from a fixation point efficiently because saccades are generated and typically completed before the larger and heavier head and neck can move (Zangemeister & Stark, 1981) and its signal has been analyzed in considerable detail using adult subjects and comprehensive calibration procedures (Baloh et al., 1975; Bergenius, 1984). EOG is also a desirable tool for measuring the VOR, since the subject's head is free to move, though this head movement introduces a problem which is discussed below. The EOG signal is undesirably altered when light levels change and when the subject blinks, grimaces, or otherwise moves facial muscles. In addition to the problems caused by blinks, the lid movements made during vertical eye movements contaminate the EOG signal. The resting potential, obtained when the eye is not moving, can also drift over time (Maurer, 1975b), requiring within subject calibration checks. Electrodes must be placed precisely or (a) any increased distance of the electrode from the desired site decreases the strength of the signal, and (b) incorrect placement increases the probability that crosstalk from vertical or torsional eye movements will contaminate the horizontal signal. Finally, the difficulty of placing electrodes on the inner canthi, especially with infants, makes individual simultaneous recording of both eyes almost impossible. Becker (1989) summarized the use of EOG:

> "(it) has been for many years the most popular [recording technique] because of its easy application and its wide range of (approximate)

> linearity. Unfortunately, it is in many respects also the least trustworthy method . . . [but it] remains indispensable for the investigation of large eye movements (in particular during combined eye-head movements), for fast screening, and for work with patients. However, data obtained by EOG may have to be corrected by empirical adjustment factors in order to exactly match those produced by other methods, and for small saccades or when monocular recording is attempted they may be definitely wrong." (p. 18,19)

Corneal reflection system - CR systems have been utilized for eye movement and pupillometry studies since the 1960's. The essential difference between the CR systems and EOG is that the CR systems *measure the eye position relative to a fixed light source*, or sources. Several CR systems have been painstakingly implemented in studies which measure infants' scanning eye movements (Bronson, 1982, 1983, 1990a; Haith, 1969, 1980; Salapatek & Kessen, 1966). An automated CR system from Applied Science Labs (ASL) has been used to measure the scanning, saccadic tracking and localization and pursuit eye movements of infants. A summary description of this latter system can be found in Young and Sheena (1975), but it has been modified for use with infants (i.e., Aslin 1981, 1985: Hainline 1981a, b: Shea & Aslin, 1990). Aslin (1981), Bronson (1982) and Hainline (1981a, b) introduced the use of online video information which provided eye position co-ordinates at the rate of 30 or 60 Hz, a significant improvement over the 6 Hz sampling rate available from previous film systems (i.e., Salapatek & Kessen, 1966). The ASL system has a linear range of +/-15 to 20 deg from a central fixation point (Young & Sheena, 1975), horizontally and vertically.

The ASL system has been used in much of the infant saccadic and smooth pursuit work discussed below. One strategy has an adult experimenter holding the infant so the infant is looking over the shoulder of the experimenter (Hainline, 1981a, b; Hainline et al., 1984; Harris et al., 1988). The infant sees the image of a stimulus presented on a TV monitor and viewed through a "hot" mirror. Simultaneously, the infra-red sensitive TV camera shoots the image of one of the infant's eyes reflected off the front surface of the mirror. Alternatively, the infant is held in the lap of an experienced holder and the infant views the image of a stimulus presented on a TV monitor (Aslin, 1985; Shea & Aslin, 1990). This image is reflected off a "cold" mirror. Simultaneously, the infra-red sensitive TV camera shoots through the mirror to view one of the infant's eyes. In both cases the camera uses a closeup lens to provide an image of the infant's pupil and the camera's field of view is approximately 2-3 cm laterally and vertically. In both designs, the holder of the infant views a monitor which shows the camera's view of the infant's eye, and the holder strives to keep the infant's eye within the camera's field of view for each trial. An infrared light source is aligned and collimated with the center of the camera lens to create a white pupil from the fundus reflex and an even whiter corneal reflection. The size of the pupil and the relative location of the corneal reflection to the centroid of the pupil are analyzed in the ASL system and produce a 60Hz measurement of pupil size and x-y location of the eye.

Problems - both EOG and CR systems have their limitations and problems. However, if one is aware of these problems and controls for them, and, further, if one asks questions within the capabilities of the system, then *both* procedures will provide reliable and accurate records of infant eye movements. The most frequent problems cited are (a) invasiveness and head restraint, (b) head movements, (c) temporal resolution, (d) spatial resolution, and (e) calibration.

The first problem concerns the invasiveness of the two systems. EOG requires the attachment of surface electrodes near the infants' eyes. Before attachment, the skin must be rubbed clean of dead cells and dirt, and the electrodes fastened with cream and tape. While practitioners of the art consider this procedure routine, special materials must be used to attach the electrodes, and special care taken to remove them for fear of irritating or injuring the young skin (Shupert & Fuchs, 1988). In addition, the older the infant, the more vigorous the infant's attempts to grab the electrode wires and remove the electrodes. With practice, though, experimenters avoid skin irritation and forestall undesired removal of the electrodes until data have been collected. In contrast, the ASL CR system does not require any attachments to the infant. It does require that the infant's head be held relatively still within the 2-3 cm field of view of the camera. While this may seem difficult, like many other procedures, including electrode placement for visual evoked potentials (VEPs) and EOG, it becomes easier with practice. As with those methodologies, experimenters do not make any pretences about being able to collect a full data set from *every* infant on *every* visit, but if the infant is awake, it is extraordinary if some data are not collected. These sessions frequently involve "clucking" or making other noises, using pacifiers liberally, and taking strategic breaks, but especially with younger infants (under 10 weeks), success rates are usually over 80%. Even with infants patched for monocular testing (i.e., Shea, Doussard-Roosevelt & Aslin, 1985), the "hit rate" is still around 70%. For example, Ornitz et al. (1979) tested subjects using EOG between 1 month and 11 years in a vestibular nystagmus task and actually had a higher percentage of successful trials for subjects under 8 months than they did for subjects over 4 years. The percentage of successful subjects was essentially the same for infants under 5 months as children 4 to 6 years. Younger infants cannot support their own heads and it is unusual for them to object to someone providing support. Older infants, however, do begin to object. Again, Ornitz et al. report half the number of successful infants at 8-10 months as they do at less than 5 months.

The second potential problem with both systems is head movements. Because EOG measures eye position *relative* to head position, any head movement changes the EOG signal. If the eyes are fixed on some point, a lateral head movement on 1-2 cm can cause a signal change of 10° (Aslin, 1985). If the corneal-retinal potential is being measured and the head position is not, then the signal change could be incorrectly interpreted as an eye movement, thus introducing a large source of error. As a result, experimenters using EOG must either record head movements and then subtract them from the eye movement signal (Regal, Ashmead & Salapatek, 1983; Tronick & Clanton, 1971), or they must eliminate them by holding the subject's head still (i.e., Finocchio et al., 1990). As mentioned earlier, this problem is not sufficient to preclude the use of EOG, but it must be controlled.

The CR systems require that the infant's eye be held within the field of view of the camera, but the head need not be absolutely still. Because the system calculates the location of the eye based on the relation of the corneal reflection to the pupil, the image of the eye can be anywhere on the screen and the ASL system will still give an accurate measurement. In contrast to the potential 10° error introduced by an uncorrected 2 cm movement in EOG, this 2-3 cm range of motion in the ASL CR system allows small position shifts without fear of confounding the signal. Perhaps the most practical motion allowable by CR, other than routine small adjustments on the holder's part as the session goes on, is that it allows the infant to suck. Whether the infant is using its own fingers, somebody else's fingers, or a pacifier, the infant can suck at any time and any rate, and the ASL system will provide a clear signal of eye movement position unconfounded by the bouncing head movements which accompany active sucking. Moreover, if the subject is fixating a point while moving his or her head, the ASL system will correctly determine that fixation has not changed. This is in direct contrast to EOG, which specifies eye position with respect to head position. In the case where the head is moving while fixation is maintained, EOG tracings will indicate a gaze change where none has taken place. It is worth repeating that the permissible head movement in the ASL CR system prevents it from being used in experiments which require large head movements or rotation of the subject, i.e. VOR studies, unless the CR system is rotated with the subject. However, EOG can be efficient and reliable in VOR research. Again, it is a question of fitting the equipment to the experimental question.

The third potential problem is sampling rate. While the EOG systems can generate high sampling rates, the current CR systems are based on video signals so their temporal resolution is restricted to 60Hz, or one sample per 16.67 msec. This rate is slow when compared to an electromagnetic search coil system or a Purkinje image tracker, since these systems sample as close to a continuous rate as the collecting hardware will allow, with rates of 1000Hz common. These systems can accurately and reliably reveal acceleration, peak velocity and deceleration of saccades. There is no doubt that the video-based CR systems cannot perform such tasks (Shupert & Fuchs, 1988), and, indeed would not approach this exquisite temporal resolution even with the two- or three-fold increase in sampling rate which will probably be available in the near future. However, users of the CR systems are acutely aware that the existing sampling rate provides a single sample only every 17 msec and comparing samples so far apart within a saccade which may last 70 or 100 msec in an attempt to map out velocity changes would be overstepping the limits of the system. However, this sampling rate is adequate to detect the existence of saccades, as well as their direction and magnitude, *if* the user calibrates the system, and *if* the saccades are within +/-15 to 20 deg of visual angle. Within the proper experimental context and question this sampling rate is not a problem. If the question is: did a subject make a saccade within a 500 msec window of time, then 60Hz is certainly fast enough to provide an answer. If, however, the question is: did the subject show open loop gain[4] (or drift) in some 30 msec window, then obviously a system which is only sampling every 16.7 msec cannot provide an answer. Indeed, some authors feel any sampling rate under 200Hz is inadequate for computation of saccadic velocities (Juhola, Jantti & Pyykko, 1985). For the same reason, if the question is: how does the velocity of a saccade change over its 100 msec duration, then the 60Hz rate is simply too slow to provide an answer. The ASL system can indicate that a saccade has

[4] Gain is expressed as a ratio: (magnitude of output)/(magnitude of input). In the case of velocity gain of smooth pursuit, gain = (the velocity of the eye movement)/(the velocity of the target). Thus, perfect following of the target by the subject would yield a gain of 1.0. A subject showing 20° /sec eye movements when a target is moving at 30° /sec would have a gain of .67.

taken place. Within its spatial resolution, it can indicate the saccadic direction and magnitude. Within its temporal resolution, it can indicate the saccadic duration. It can give an estimate of the average saccadic velocity over the 100 msec duration, or an estimate of a corrected peak velocity (see esp. Harris, Abramov & Hainline, 1984 for details of the construction of nomograms for use in these analyses). It cannot provide detailed information about changes in velocity within the saccade from one millisecond to the next.

Fourth, the spatial resolution of EOG is usually estimated at 1° (Finnocchio, Preston & Fuchs, 1990; Woestenburg, Verbaten & Slangen, 1984; Young & Sheena, 1975), and, under some testing conditions, .5° (Larson, 1970). The ASL CR system's optimum resolution is about 15 min of arc. With adults there is the added advantage of presenting small targets, i.e. 15 min of arc, for fixation and localization, so the system can be calibrated to its maximum level. Infants will not reliably fixate a target much smaller than 1 deg so any given infant's response can only be calibrated to within +/- 0.5 deg. With larger targets, this expands to 3 to 4 deg (Hainline & Lemerise, 1985). While this resolution would cause problems measuring the fine detail of saccadic localization in trained adults, this resolution is sufficient for testing infants and for asking many questions about the adult oculomotor system (i.e., Aslin & Shea, 1987). The level of accuracy obtainable with the ASL system is sufficient especially when one considers the problems involved with the calibration of infant eye movements.

This problem, which is the fifth and last potential problem of any eye movement system, is calibration. This critical topic has been discussed in great detail for all systems used to measure eye movements in infants (Aslin, 1985; Bronson, 1983, 1990a; Finocchio et al., 1990; Harris, Hainline & Abramov, 1981; Metz, 1984; Salapatek et al., 1972; Sheena & Borah, 1981; Shupert & Fuchs, 1988; Slater & Findlay, 1972a, b, 1975b; Woestenburg, et al., 1984). When an adult is tested, the experimenter defines a calibration pattern and indicates the desired fixation points. The adult fixates the points in the sequence and duration the experimenter requires. Because this procedure does not work with infants, modified techniques have been developed which move a small target within the stimulus field, stop it at several locations, and collect eye position data after the infant has time to catch up to the stimulus. Even so, one cannot instruct an infant to fixate a particular target at a particular time, and their best fixations to stationary targets are more variable than adults (see, for example, Haith, 1980; Harris et al., 1988). Harris et al. (1981) point out that while the infant's fixation of a small target may consist of a large range of eye positions, an appropriate statistical analysis will yield an accurate estimate of the infant's line of sight (see also Bronson, 1982; Hainline et al., 1984). Eye movement tracings gathered during the calibration trials are inspected visually by the experimenter after each trial or off-line after the completion of the session to ensure that noise such as blinks, or off field viewing is not included with calibration data. This visual inspection is accompanied by software algorithms which maintain constant standards of acceptable data. It is not always possible to gather as much calibration data from an infant as from an adult. Infants have such a short attention span one could spend so much time collecting calibration data the infant will be asleep or too restless to collect experimental data. Nevertheless, careful calibration of EOG and CR systems yield a reliable estimate of infant eye movements, as well as a baseline upon which to make between-subjects comparisons (Aslin, 1981; Bronson, 1982; Harris et al., 1981).

In conclusion, the ASL CR system and EOG both provide accurate information about the direction, magnitude and latency of saccadic and smooth pusuit eye movements. These systems are not as accurate, either temporally or spatially, as a magnetic search coil system or an SRI Purkinje system, but since neither of those systems can be used with infants, we return to the point that if the experimenter matches the questions being asked in the experiment to those answerable by the technology, and subsequently couches the interpretation within those restrictions, the level of accuracy is sufficient. Whether a CR system or EOG is used should be determined by the parameters of the experimental question and the unique strengths and weaknesses of the two measurement techniques.

B. Working with infants in general

Anyone who has tested infants, or tried to test infants, immediately embraces the masterful understatement that "(the) labile alertness of infants places a premium on finding the most seductive stimuli" (Shupert & Fuchs, 1988, p. 587) and the comment that investigators interested in infant vision "have encountered formidable difficulties posed by labile attention and limited response repertoire" (Fox et al., 1980, p. 323). Reseachers who work with adults expect that their subjects can be trained to a certain paradigm, will provide some reliable response for a predictable period of time (even across days or months), and will listen to instructions. Researchers who work with non-humans expect that their subjects can be trained and, once trained, will respond reliably across time. In addition, in most cases, they will work for food or liquid reinforcement. None of that works with human infants. None of it. Add to those problems that young infants sleep at irregular intervals, eat at unpredictable times and generally respond only to their own internal instructions. Older infants, while more predictable about eating and sleeping, quickly reach an age where they do not like having their heads held for very long, do not like dark rooms and will not tolerate being held for long by a stranger. To salvage this situation, experimenters must rely heavily on another staunch characteristic of awake infants: they like to look at things. More importantly, they like to look at new things. Experimenters bank on their creativity in being able to find those "seductive stimuli", construct testing paradigms that will elicit infants' responses in the shortest amount of time, and prepare to wait out the infants' non-scientific behavior.

The perceived difficulty of the tasks and tricks involved in the description above varies widely with what the experimenter is familiar. For example, researchers who use EOG tend to feel that applying recording surface electrodes is easy, yet holding an infant still for CR systems is difficult and eliminates many infants as unco-operative (i.e., Shupert & Fuchs, 1988). Researchers who use CR would respond that most infants do tolerate it for the necessary period of time. They would also argue that the infants tolerate the head restraint better than they would tolerate the "invasive" electrodes necessary for EOG (Aslin, 1981), and that for most EOG procedures the head needs to be held still anyway. In fact, younger infants are easier to test than older infants simply because they tolerate the restraint so readily - possibly because it provides them with head support they do not yet have themselves. Researchers who use CR, most of whom also have experience with EOG, would quail at the prospect of returning to that procedure unless the experimental question demanded it. The electrodes used in EOG are not always easy to apply, they do not always stay in place once they are applied, and some infants react badly to the rubbing involved with cleaning the electrode site and/or to the cream involved. EOG researchers would immediately rebut that is not the case, that the electrodes are trivial and few infants are bothered by the procedure. And the debate would continue. The strategies and preferences cannot be resolved here. The crucial point is that *all* techniques must be appreciated both for their strengths and weaknesses, and that the experimenter must understand both, and maximize the former while minimizing the latter. If this is done, then *both* systems can contribute to our understanding of oculomotor development. It may help to keep in mind here that while it's true that not all infants will yield acceptable data, an oculomotor task which uses nonhuman primates can require *months* of training and even then not all animals are good performers. To a much smaller degree, even naive human adults fail as subjects if they can't restrain blinking or if they can't maintain fixation. Again, if one concentrates within the restrictions of the paradigm a priori, this is not so much a problem as it is part of the job.

4. Elicited eye movements in human infants

A. Types of eye movements

The types of eye movements discussed in this chapter are: vergence, smooth pursuit, vestibular and saccadic. Optokinetic nystagmus (OKN), frequently classified and discussed as a separate type of eye movement (Buttner & Buttner-Ennever, 1988), will be discussed here under smooth pursuit and vestibular eye movements. In each section, the eye movements will be defined, anatomical units described and behavioral or electrophysiological research presented.

B. Vergence

Vergence eye movements are rotations of the eyes toward (convergence) or away from (divergence) each other to maintain bifoveal fixation of a target or a point in space. In adults, these eye movements are initiated by retinal disparity cues (fusion vergence), and/or by tonic, accommodative (blur) or proximal stimuli (accommodative vergence) (Buttner & Buttner-Ennever, 1988). Ordinarily these kinds of information occur simultaneously, that is, an object moving towards a subject generates both changing disparity and blur. Also in adults, failure to maintain bifoveal fixation results either in diplopia (double images) or in suppression of the information from one of the two eyes.

Vergence eye movements are onc third of the "near triad", the other two components being accommmodation and pupil size. These three motor responses are intimately linked in adults. While the vergence and accommodative capabilities of infants are being researched in detail, their synkinesis within the triad is relatively unexplored (for a current review, see Aslin, 1992), although it has been shown that, as in adults, vergence and accommodation become uncoupled when young infants are placed in the dark (Aslin & Dobson, 1983). The third member of the triad, pupil size, has received only slight attention, and then sometimes as a dependent measure and not a research question itself (Munsinger & Banks, 1974; Salapatek, Bechtold & Bergman, 1977; Birch & Held, 1983; Isenberg, Molarte &

Vazquez, 1990; Robinson & Fielder, 1990, Shea et al., 1985; see also Table 2). Vergence movements are generally slow, with a maximum velocity of about 20° /sec, a duration approaching 1 sec, and a latency to respond of approximately 160 msec, but saccadic vergence responses do exist (Barbeito, Tam & Ono, 1986; Hain & Zee, 1989). Whatever their speed, they can be either symmetric or asymmetric, depending on whether the target is moving directly on the midline, or off to one side (Fig 1 D and E).

As noted earlier, vergence is an odd eye movement in that the muscle pairs which act in unison in saccades or pursuit, are "broken" in vergence. For example, when saccading horizontally, the OS lateral recti and the OD medial recti contract as a pair, and their antagonists, the OS medial recti and the OD lateral recti, relax as a pair. In vergence the pairs are redefined: in *di*vergence the lateral recti contract as a pair, and the medial recti relax as a pair, while in *con*vergence the reverse occurs. The efferent pathway for vergence eye movements apparently starts in the striate, parietal, medial temporal (MT) and frontal cortex. Signals traverse oculomotor interneurons, abducens nuclei and premotor nuclei in the midbrain enroute to the oculomotor nuclei and the medial and lateral recti (Maunsell & van Essen, 1983; Hain & Zee, 1989; Nieuwenhuys, et al., 1988). While the SC, pretectum and possibly the cerebellum are involved with vergence signals, the details of the vergence pathways are not well understood and are still being researched (Buttner & Buttner-Ennever, 1988).

The convergence angle is formed by the rotated position of the eyes, so larger angles indicate closer targets. Because the eyes cease rotating outward, diverging, after a target reaches a distance of about 2 meters from an adult human, vergence only yields absolute distance information within this range (von Hofsten, 1976). Since the young infant spends so much time investigating close things like hands, feet, bottles and parents' faces, vergence could potentially provide considerable information about objects moving in near space. Support for this argument comes from research on the resting position of vergence. While the average resting position of adult vergence is between 80 and 120cm (depending on the measurement technique used, see Owens & Leibowitz, 1980; Aslin, 1987a), the average resting position for 1 to 5 month old infants is 31cm (Aslin & Jackson, 1981). Between 5 and 12 months, this position shifts to 51cm, and by 18 to 26 months, to 48cm (Aslin, 1986).

Early observations indicated that newborn infants could neither move their eyes conjugately nor binocularly fixate a target (Gessell, Ilg & Bullis, 1949;

Table 2

Motor Development "Landmarks" Before 5 Months of Age

Motor Ability	Description	Age	Source
control of head & neck	turn to one side **turn to both sides** chin held off contact surface **good prone control**	birth **1st week** 2 m **3 m**	Gallahue, 1989
control of trunk	lifts head and chest **attempts supine-to prone position**	2 m **3 m**	Gallahue, 1989
sit	sits with support	3 m	Gallahue, 1989
reflexive pull-up reaction	attempts to maintain upright position when tipped backwards	3 or 4 m	Gallahue, 1989
forward and downward parachute reactions	limbs move toward ground when tipped forward or lowered	4 m	Gallahue, 1989
labyrinthe righting reflex	infant attempts to maintain upright head position when trunk is tilted from vertical	appears 2nd month: gets stronger in 1st year	Gallahue, 1989
(a)symmetrical tonic neck reflex	stereotypical extension of unilateral arm and leg when neck is moved in one direction	3- and 4-month-olds assume asymmetrical poses about 50% of the time, then decreases until 6 m	Gallahue, 1989
smooth eye movements	OKN & vestibular	birth	Telang & Dweck, 1989
smooth eye movements	pursuit tracking	appears at 6-8 w, but only at a particular velocity - accuracy increases rapidly	Aslin, 1981 Shea & Aslin, 1990

Table 2 (continued)

saccadic eye movements	localization	saccades to localize targets at 1 m	Aslin & Salapatek, 1975
vergence eye movements	binocularly fixate stationary & moving targets	some evidence at 3 m - rapid improvement to 6th	Aslin, 1977
pupillary light reflex	pupil dilates or constricts when bright light added or removed	appears at 30 weeks gestation; well developed around 1 m postnatal	Teland & Dweck, 1989; Robinson & Fielder, 1990; Isenberg et al., 1990
accommodation	focus accurately to targets at different distances	4 m	Haynes et al., 1965; Banks, 1980a, b; Howland, 1982
motion perception	detect changes in moving targets based on un-confounded motion cues	rapid improvement around 3-4 m	Freedland & Dannemiller, 1987; Dannemiller & Freedland, 1989; Aslin & Shea, 1990
disparity processing	realign eyes (change vergence angle) in presence of large disparity cue	4 m	Aslin, 1977
stereoscopic vision, including stereoacuity	respond to targets in stereoscopic space with only cue binocular disparity	rapid improvement between 3.5 and 4.5 m	Fox et al.,1980; Shea et al.,1980; Held et al., 1980; Petrig et al., 1981
binocular vision	process and respond to binocular information not necessarily stereoscopic	rapid improvement around 3-4 m	Birch & Held, 1983; Shea et al., 1987; Mitkin & Orestova, 1988

Peiper, 1963) but 2 and 3 month olds could move their eyes reliably under conditions which would stimulate a change in vergence (Ling, 1942). Later, more quantitative research triggered a controversy of methodology and interpretation (Hershenson, 1964; Dayton & Jones, 1964a,b; Dayton et al., 1964; Wickelgren, 1967, 1969; Rethy, 1969; Salapatek et al., 1972; Slater & Findlay, 1972a, b; Slater & Findlay, 1975a, b; Maurer, 1975a, b; Aslin, 1977). Without discussing all the methodological complications, the gist of the problem centered on the question of where the infants were actually sighting - along the visual axis, or along the optic axis. The difference between the two is approximately 5° in adults, and as much as 8° to 10° in infants, which would introduce a considerable error into calculations of eye position and vergence angle. Slater and Findlay (1975a) solved the problem by calculating the vergence angle as a function of target distance, and reported that newborns would alter vergence angles to targets at 50 and 25cm, but would not come any closer. Aslin (1977) solved the problem by using the target to generate corneal reflections rather than fixed marker lights (thus generating an unchanging relation between each eye's corneal reflection and the two pupil centers) and found that a target which moved between the distances of 12 and 50cm from infants under 2 months of age did not elicit appropriate changes of vergence angle.

Some of this lack of response could be attributed to infants' lack of interest in the stimuli or general inattention, at which point the vergence angle would revert to the resting position (Aslin, 1986), but the response is reliable. It is tempting to assign the reason for this inaccurate vergence in young infants to their still inaccurate accommodation, or a non-functioning tie between these two responses, but Aslin and Jackson (1979) showed that by 2 months of age accommodation and vergence are functionally linked and reliable. Another reason for infants' failure to converge and diverge accurately at young ages may be that they are not processing the retinal disparity cue to depth accurately. Aslin (1977) tested this by placing prisms in front of one eye of 3 to 6 month old infants as well as adults. Adults would generate a refixation eye movement to even a 1 diopter prism so the target's image, even though displaced by the prisms, would remain binocularly foveated. Infants did not make this response until 4 months of age, and then required 10 diopters of prism before they would respond.

Mitkin and Orestova (1988) recorded vergence movements using EOG in 20 infants between 4 and 33 weeks of age. Targets were moved between 15 and 45cm from the infants either along the midline or asymmetrically positioned in front of one eye. Between 4 and 5 months of age infants showed a rapid improvement in their co-ordinated responses to these targets. This improvement was the same whether the target was midline or in front of one eye.

The current standing of the research indicates that while directionally appropriate vergence can be observed in very young infants, but rarely in neonates (Wickelgren, 1967, 1969), it does not become a reliable response until after 2 or 3 months of age. By 3 to 4 months of age the vergence changes are also magnitude appropriate and infants can respond well to varying speeds of target motion.

C. Smooth pursuit

In adults, smooth pursuit eye movements are slow, generally about 30 to 40° /sec, with a reaction time to a moving stimulus of approximately 125 msec, though smooth pursuit has been recorded at velocities of 100° /sec (Meyer, Lasker & Robinson, 1985). The eyes move conjugately if the target is moving at a constant distance from the subject, and at a speed that matches the velocity of the moving target. If the target exceeds the maximum velocity of the pursuit system, catch-up or corrective saccades will become interspersed with the pursuit (Fuchs, 1971; Keller & Johnsen, 1990). Unlike saccades, smooth pursuit usually cannot be generated in the absence of a stimulus and has not been found in species without foveas. Smooth pursuit serves to maintain foveal fixation of a slowly moving target, but this fixation comes at the cost of blurring the background. This presents an interesting problem for the infant's visual system, especially if the infant is "interested" in things that move: if an infant is fixating an object, say a person standing 1 meter away against a patterned background, like wallpaper, when that person starts to walk and the infant starts to smoothly pursue that person as a target, smooth pursuit of that target reduces movement of the target relative to the retina, but it *increases* the movement of the background relative to the retina. If the infant is "interested" in motion, he or she might now switch to looking at the background rather than the target, because the retinal motion of the background is now greater than the target's.

Once a researcher finds a stimulus condition that can elicit smooth pursuit, and many stimulus characteristics influence pursuit (Fetter & Buettner, 1990), the researcher also finds that pursuit can be inaccurate in several different ways. First, the pursuit can be heavily interspersed with saccades. Second, the subject can lag behind the target while tracking. As Aslin (1981) points out, there are three ways this can happen (a) the amplitude of the eye excursion can lag behind the amplitude of the target excursion, causing low *amplitude* gain; (b) the velocity of the eye movement can lag behind the velocity of the target, causing low *velocity* gain; or (c) the eye can match the target velocity and excursion, but lag behind the target as it moves, resulting in a temporal phase lag. All of these inaccuracies show up in young infants when they are presented with smoothly moving targets.

Smooth pursuit eye movements involve many structures and pathways in the oculomotor system, but just how they work and exactly what "routes" are involved are still being researched (see Eckmiller, 1987, for review). The cortical area MT receives information from the visual cortex, and projects to the medial superior temporal area (MST). Cells in this area in the monkey respond specifically while smooth pursuit is being generated (Maunsell & van Essen, 1983). MST projects to the posterior parietal cortex where neurons respond differentially during smooth pursuit as targets change direction (Leichnetz & Goldberg, 1988). These areas of cortex, as well as the frontal eye fields, project to nuclei in the pons and to the cerebellum. Cells in both sites fire during smooth pursuit. The cerebellar sites are also connected to the vestibular nuclei which, in turn, are heavily connected to the ocular motoneurons. If the cerebellum is removed, smooth pursuit disappears (Bender, 1980; Buttner & Buttner-Ennever, 1988). Another pair of structures, the dentate nucleus (lateral cerebellum) and the y-group (in the brainstem with the vestibular nuclei), project to the oculomotor nucleus and have been identified as generating smooth vertical eye movements (Chubb & Fuchs, 1981). Optokinetic nystagmus (OKN) does not have the cortical connections of smooth pursuit, but instead shares some pathways with the vestibular system (Precht, 1981; also, see next section).

Research on smooth eye movements in human infants, either as smooth pursuit tracking or as the slow component of VOR or OKN, have been conducted since the early work of Ling (1942). A subject who is tracking a target with smooth pursuit will cease tracking shortly after the target moves beyond the subject's visual field[5]. OKN is a combination of smooth pursuit (slow phase) and saccades (fast phase) which occurs as a compensatory eye movement when the entire visual field moves. A common visual stimulus used to evoke OKN is black and white bars drifting in a direction orthogonal

[5] Adult studies have measured smooth pursuit in response to a target which disappears and later re-appears (Bahill & McDonald, 1983; Becker & Fuchs, 1985). In these cases, adults will generate predictive pursuit for short periods of time. Generally speaking, however, subjects will cease to track a target smoothly when it disappears from view.

to their orientation, that is, vertically oriented bars would drift left or right. In this case, the subject smoothly tracks a portion of the moving field but then makes a saccade *against* the direction of the stimulus motion, returning the line of sight to a more central position.

OKN in infants has been studied for decades, beginning with the work of McGinnis (1930) who found that even neonates showed some indication of OKN and saccadic tracking. Though the measurement was rough, McGinnis reported the consistency improved through the first months of life. The presence of OKN at such a young age allowed investigators to use it as a dependent measurement for investigating other visual abilities, i.e. visual acuity (Dayton et al., 1964a). Rather than using OKN as a measuring tool, Krementizer et al. (1979) measured OKN itself using EOG and found that, like adults, infants under 3 days of age showed OKN to a particular stimulus pattern of black and white stripes. However, unlike adults who showed increased slow phase velocity of OKN at target speeds up to 40° /sec, infants ceased showing OKN when the visual field's speed increased beyond 25° /sec. This peak velocity could be constrained by sensory or motor capabilities, or, again, by a combination of both.

Harris, Cassel and Bamborough (1974) presented 8 to 24 week old infants with objects which moved at a rate of 9°/sec against a stationary background, or objects that moved with the background, or objects that moved in a direction opposite to the background. Observers viewed the infants' eyes and recorded whether the infant followed and fixated the target. Harris et al. concluded tracking was worse in the latter case, that is, that relative motion is important for infants to be able to generate tracking.

Aslin (1981) provided a thorough description of the development of smooth pursuit in a longitudinal study of young infants. With the ASL CR system described above recording their eye movements, infants were presented with a high contrast target which oscillated horizontally in a sinusoidal pattern across a range of 20 deg at 5 velocities between 10 and 40 deg/sec. The very young infants responded with the eye movements shown in Fig. 2A. The "staircase" pattern of tracking indicates a series of saccades and inter-saccadic fixation intervals. There is little, if any, smooth pursuit. The saccades are presumably catch-up saccades initiated by an infant who has fallen behind the target. They are similar to those generated by adults who are presented with a target which is moving too quickly to be pursued smoothly. The first reliable evidence of smooth pursuit appeared in 6 to 8 week old infants. When it first appeared it was only in short segments and with a poor velocity

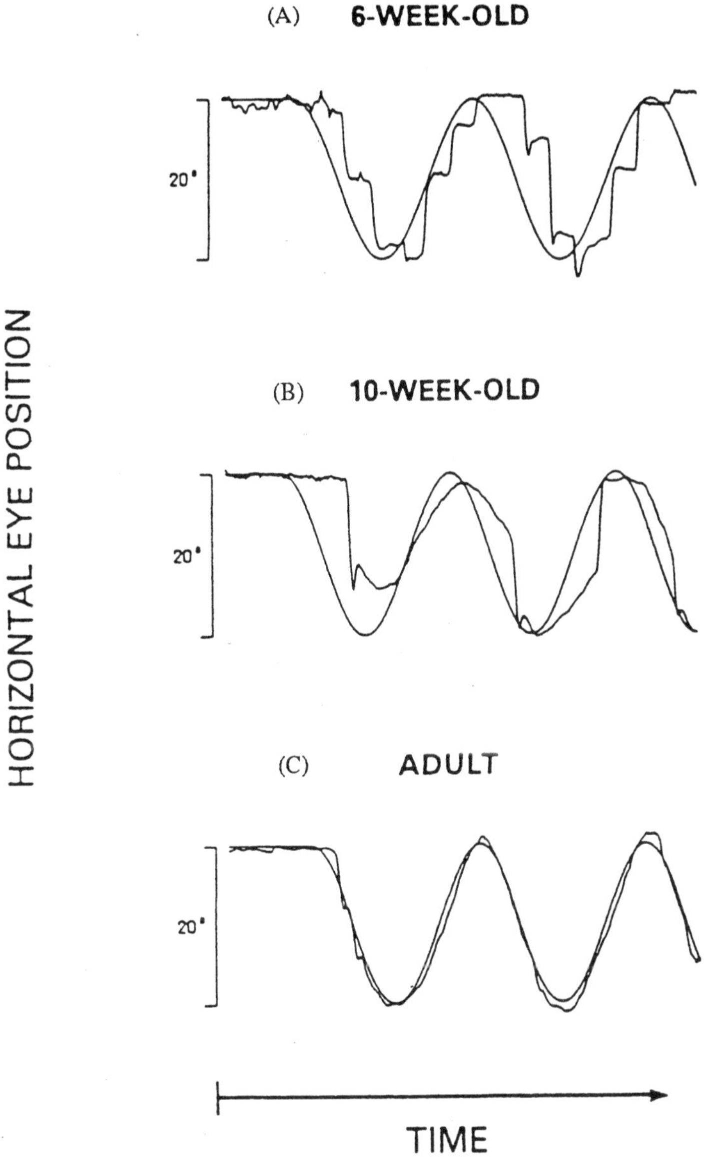

Fig. 2. Sample pursuit eye movement tracings from (A) a 6-week-old infant, (B) a 10-week-old infant, and (C) adult. X-axis is time, Y-axis is magnitude of target motion. All tracings obtained with corneal reflection system sampling at 60 Hz. See text for details. Target was moving horizontally, so upward deflections are rightward eye movements. Reprinted from Aslin, 1987c, with permission.

gain of approximately .50. By 12 weeks of age representative tracking responses looked like those in Fig. 2B. Saccades are still present, but occur infrequently, and patches of smooth pursuit occur. The initial eye movement in these trials was almost always a catch-up saccade with an average latency between 500 and 800 msec, far higher than the standard latency for smooth pursuit. An adult's response is shown in Fig. 2C. Aslin also found that tracking at the low target velocities was more accurate than at the high velocities as well as some evidence of predictive tracking in the older infants. Shea and Aslin (1984) extended this work by measuring vertical tracking as well as horizontal and also succeeded in testing some very "old" infants. Fig. 3A shows the eye movement responses of an 11-week-old infant tracking a target moving vertically. Note that while by this age infants are showing smooth pursuit when tracking horizontally, this tracing is almost entirely saccadic. In Fig. 3B, a 13-week-old infant tracking a vertical target shows segments of smooth pursuit, generally in the downward direction. By 37 weeks (Fig. 3C), the infant is showing long stretches of smooth pursuit with few saccadic intrusions. Overall, infants showed vertical smooth pursuit at a later age than horizontal, and were consistently less accurate and showed lower gain when tracking vertically, a finding not qualitatively dissimilar to the poorer vertical smooth pursuit found in many adults (Collewijn & Tamminga, 1984; Baloh et al., 1988).

The saccadic tracking of the young infants is interesting because the infants are consistently on the target's path, they're just not using smooth pursuit. They clearly see the target, detect that it's changing position, and respond to it, but either they're (1) not accurately processing the changing position and so can't generate smooth pursuit to it; (2) not perceiving the target motion as smooth; (3) not capable of generating a smooth pursuit eye movement; or (4) not capable of generating a smooth pursuit eye movement fast enough to keep up with the target. Recent research has indicated that infants' motion perception is sufficient to detect this moving target (Freedland & Dannemiller, 1987; Dannemiller & Freedland, 1989; Aslin & Shea, 1990), but detection is not the same as integrating this motion signal into the smooth pursuit system and eliciting an efficient response. Smooth eye movements have been recorded in newborns but they have been VOR or OKN movements, not smooth pursuit to a small target. These oculomotor responses are similar, but the stimuli that elicited them, the anatomical systems that underlie them, and perhaps other aspects, are not.

Roucoux, Cullee and Roucoux (1983) tested the hypothesis that young infants would not generate smooth pursuit because their immature fovea render

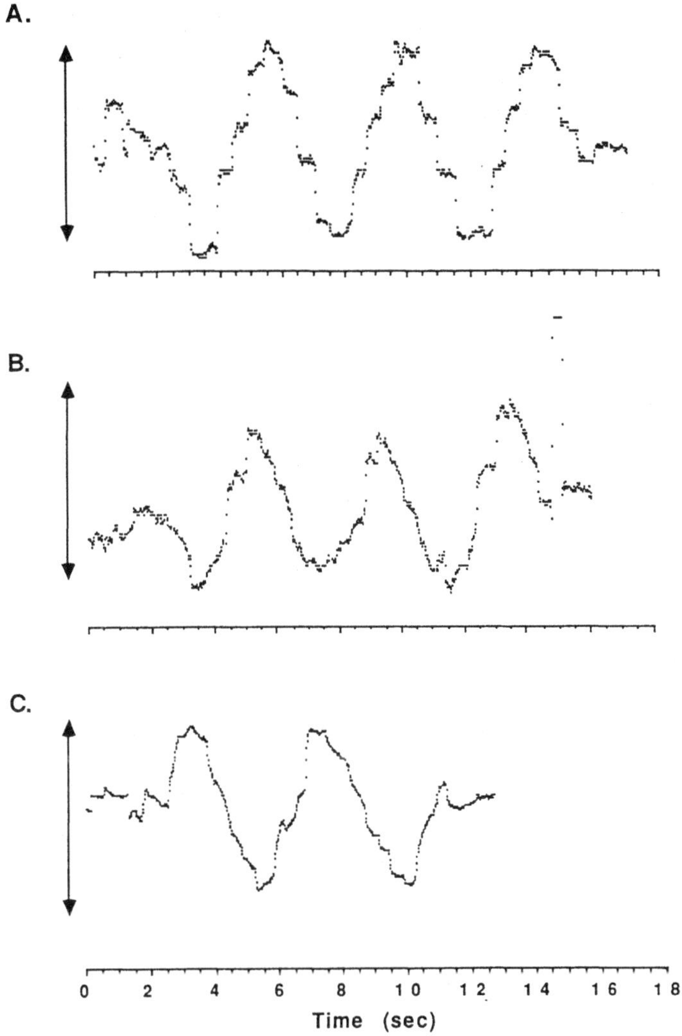

Fig. 3. Eye movement tracings of 3 infants to a smoothly moving target. X-axis is time, Y-axis is magnitude of target motion. All tracings obtained with corneal reflection system sampling at 60 HZ, began and ended with a 2 sec buffer obtained with an unmoving, blinking central fixation point, and target speed was 10 deg/sec. Target was moving vertically, so an upward deflection is an upward eye movement. (A) 82-day-old infant. (B) 94-day-old infant. (C) 258-day old infant. Data from Shea & Aslin, 1984.

them "comparable to . . afoveate animals" and that only as the fovea matured would they be able to generate smooth pursuit. Roucoux et al. tested 6 infants using EOG and allowing the head to move freely. Targets were black and white faces ranging in size from 2 to 10 deg. While no quantitative analyses are presented, Roucoux et al. reported that infants could smoothly pursue a target moving at 10 deg/sec at 4 weeks of age, though the velocity of the pursuit, and thus the gain, was poor. Faster target velocities tended to elicit saccadic responses and head movements.

Shea and Aslin (1990) further explored the apparent immaturity of the infants' smooth pursuit capabilities. Instead of utilizing the sinusoidally oscillating targets of earlier studies, they presented 10 infants between the ages of 7 and 11 weeks with a step-ramp target. This target was displaced instantaneously from a center fixation point (step) and then moved at some velocity for some distance (ramp). It was hoped that this step-ramp target would capture the infants' attention and, since it did not change direction once smooth motion began, it might be simpler for the infants to track. Target velocities of 3, 6 and 12 deg/sec were used for the infants, while control adults saw only 3 and 6 deg/sec. Even the youngest infant showed some periods of smooth pursuit, though there were many saccades. The gain, or efficiency, of the smooth pursuit segments is summarized in Table 3. The constancy of the infants' smooth pursuit velocity across a quadrupling of target speed was completely unexpected and intriguing. It may reflect an error in the conversion of velocity signals into smooth pursuit commands. Such uncalibrated responses do occur in other visual responses of young infants, i.e., saccades (Aslin & Salapatek, 1975) and vergence (Aslin, 1977) and suggest additional research.

There is some evidence that smooth pursuit is improving in childhood and even into adolescence (Herman, Maulucci & Styck et al., 1982; Kowler & Martins, 1982; Palmer & MacDonald, 1990), though there are disagreements about the quantity and quality of these eye movements and the role attentional problems and expectations play (for example, see Aslin & Ciuffreda, 1983; Dannemiller et al., 1983; and Kowler & Martins, 1983 for reply).

D. Vestibular eye movements

A complete review of the vestibular system is beyond the scope of this chapter, but the anatomical, physiological and functional interactions between the vestibular system and the oculomotor system warrant some

Table 3
Characteristics of Smooth Pursuit

Subject Group	Target Velocity	Average Gain	Average Eye Velocity	Peak Gain	Peak Eye Velocity
Adult	6 deg/sec	.91	5.46 deg/sec	1.07	6.42 deg/sec
	12	.76	9.12	.87	10.44
Infant	3	.503	1.51	1.91	5.73
	6	.247	1.48	.89	5.36
	12	.113	1.36	.25	3.0*

From Shea & Aslin, 1990

*artifically low because it includes those trials in which an infant did not make any eye movement at all until after the ramp portion of the trial was complete discussion. The vestibular system is involved in some of the infant's earliest reflexes. In the rooting or search reflex, for example, the newborn infant must be able to localize the touch stimulus and turn its head toward the touch (Gallahue, 1989). The time-honored strategy of rocking and cuddling infants stimulates the vestibular system and frequently soothes the infants (Ornitz, 1983). The vestibular ocular reflex (VOR) is frequently labelled as a "phylogenetically old" eye movement. The VOR is a slow eye movement which is equal to and opposite a rotatory head movement. It serves to stabilize the direction of gaze in spite of body and head motion in several axes. Like OKN, vestibular nystagmus is a two-part response, one fast and one slow, which occurs especially if the body and head motion are maintained for long periods of time. When the head moves, the visual image can blur on the retina. If, however, eye movements are generated which compensate for this head movement, the visual image can be stable.

Much of the VOR research involves rotating the subjects around the vertical axis, for example, spinning them while sitting in a chair while in the dark. Darkness is necessary because if the subject fixates, or tries to fixate, a stationary target, the VOR is facilitated. If the subject fixates a target which is moving with him/her, the VOR is suppressed. The type of rotation pattern will determine the type of vestibular response. Repeated head rotation generates *vestibular nystagmus* - slow compensatory eye movements interrupted periodically by saccades in the direction the head is rotating. A subject rapidly moved from rest to a constant velocity shows *perrotatory nystagmus* - slow compensatory eye movements with saccades in the direction the body is moving. A subject rapidly moved to rest after moving for some period of time at a constant velocity shows *postrotatory* nystagmus - the pattern is opposite of perrotatory, so saccades are against the direction the body was moving.

Recall that the three semicircular canals of the labyrinth are laid out in angles similar to the extraocular muscles to which they send excitatory signals. These excitatory signals travel the afferent vestibular nerve to the four vestibular nuclei in the brainstem (lateral [Deiters'], medial, superior, inferior - Highstein & Reisine, 1981; Nieuwenhuys et al., 1988). Those nuclei project to the abducens nuclei and send inhibitory and excitatory information directly to the motoneurons of the appropriate ipsilateral extraocular muscle pair (Buttner & Buttner-Ennever, 1988). These nuclei also respond to optokinetic information and the two systems interact under a variety of conditions (Precht, 1981; Westall, 1986). The accessory optic system, which involves three areas in the mesencephalon with projections to the InfO and from there to the flocculus, is also involved with visual-vestibular interactions.

It receives contralateral retinal input and its sensitivity to direction and velocity enables it to play a role in compensatory eye movements (Nieuwenhuys, et a., 1988), as well as possibly, smooth pursuit (Mustarti & Fuchs, 1989). The cerebellum, of course, is also involved with the VOR, since the VOR is suppressed by visual fixation and it plays a role in processing and controlling eye movements (Robinson, 1986). While the cerebellar flocculus is not the site of motor learning in the VOR, it must be present for motor learning to take place, possibly because of its visual signal output and its inhibitory synapses on the brainstem (Lisberger, 1988). The VOR itself does not have cortical input, but consists of three neurons: the vestibular nerve, the vestibular nuclei, and the oculomotor nuclei. The first two are also called the primary and secondary vestibular neurons, respectively. The co-ordination of these neurons has been much discussed, and a site proposed for the integration of their signals: the nucleus prepositus hypoglossi, also known as the prepositus nucleus, which is located adjacent to the vestibular nuclei (Baker, Evinger & McCrea, 1981). This nucleus also has cerebellar connections, and receives visual input so it is involved at a number of visual and oculomotor levels (Buttner & Buttner-Ennever, 1988; Nieuwenhuys et al., 1988).

Work on vestibular system in infants dates back at least to the work of McGraw (1941) and some of the early work is summarized by Ornitz (1983). McGraw watched the eye movements of children between birth and 2 years of age as she held them upright and rotated them with her as she pirouetted. Infants under 1 month moved their eyes and head slowly in the direction of motion while moving, and kept them there until they had stopped, at which point they made a slow movement in the opposite direction ("lateral deviation phase"). Starting around 4 months of age, infants made slow horizontal eye movements while moving, which continued briefly when they stopped ("gross oscillatory phase"). Around 7 months, the pattern shifted to rapid horizontal eye movements while moving ("fine oscillatory phase"). This latter pattern, while showing marked similarities to an adult response, was not stable in all children tested until 2 years of age. In an extension of this work, Lawrence and Feind (1953) tested infants under 10 days of age using a more sophisticated rotation device. Infants were rotated with the lateral semicircular canals in the plane of rotation at the rate of 10 times in 20 seconds, with consistent acceleration, deceleration and velocity of motion. The majority of infants showed an initial head rotation opposite to the direction of body motion and nystagmus with a fast component in the direction of rotation. As rotation continued, the nystagmus ceased. When rotation stopped, the head and eyes rotated in the direction of rotation and nystagmus showed a fast component opposite the direction of rotation. The direction of the nystagmus was constant, its rate and intensity were not.

Eviatar, Eviatar & Naray (1974) used the torsion swing and the ice-cold caloric method (ICC) and found that at 1 month, over 80% of the full term infants showed nystagmus, but none of the premature infants did. All older infants showed ocular deviation in the direction of the slow component of nystagmus. This percentage was higher than obtained with very young infants (Mitchell & Cambron, 1969) when rotation and ICC elicited nystagmus in approximately 40% of infants under 2 weeks of age. In another application of those methods, Eviatar et al. (1979) and Eviatar and Eviatar (1979) compared the nystagmus response in 276 infants who were either large-size for fullterm gestation, appropriate-size for gestation age, or small-size for gestation age. The results of the swing test showed the highest percentage of positive response in the largest babies, where "positive response" was true nystagmus with both the fast and slow component. The amplitude and slow component velocity increased from birth to 6 months in the fullterm large infants, but the frequency of the nystagmus did not change with age. For the ICC condition, the frequency, slow component velocity, amplitude and latency changed with age. The first three measures increased from birth to 9 months while the latency decreased from birth until 6 months. By 3- to 6- months all but the premature infants showed mature responses. Eviatar and Eviatar point out this is about the time primitive brainstem reflexes disappear (see also Table 2) and when the righting response appears in fullterm and regular size infants.

Ornitz et al. (1979) measured VOR in subjects between 1 month and 11 years and concluded that a typical infant response is more intense (i.e., higher amplitude, increased frequency), but more compressed in time (shorter duration) than older infants or adults. In addition, the slow component of secondary nystagmus is slower than the slow component of primary nystagmus at all ages, but the discrepancy is higher in young infants. Ornitz et al. note that this rate of change is not linear and suggest a dual sequence of maturation: (a) that the mean velocity of primary nystagmus develops gradually throughout childhood, mirroring the development of vestibular responsiveness, and (b) the slow component of secondary nystagmus matures basically in the first 30 months, showing the development of vestibular adaptation. Ornitz, Kaplan and Westlake (1985) measured the gain and time constant of the VOR to constant acceleration in complete darkness in 2 to 130 month old infants and control adults. The time constant of primary perrotatory nystagmus at 5 months is about 7.5 sec and 10.5 sec in adults. The greatest increase in the time constant occurs between the 5th and 10th month and may be related to inhibition developing in the brainstem (Ornitz et al, 1985; Eviatar et al., 1979). Other researchers have also found increased VOR gain to both velocity pulse and sinusoidal oscillations, as well as shorter time constants in young infants (Goodkin, 1980; Regal et al., 1983; Weissman, Leigh & Discenna, 1986; Ornitz & Honrubia, 1988; Weissmann, Discenna & Leigh, 1989; Finocchio, Preston & Fuchs, 1991). Additional work has indicated that children as old as 14 or 16 years still have an elevated VOR intensity (Herman et al., 1982; Ornitz, 1983).

Reisman and Anderson (1989) measured eye movements in 2 and 4 month old infants who were rotated under light and dark conditions at .10, .25 and .50 Hz. The VOR phase was nearly adult-like, but gain ranged from .40 to 1.40. The cervico-ocular reflex (COR), which involves muscle and joint receptors in the neck and is obtained when the body is rotated and the head held still, had a similar phase, but a significantly higher gain and more intrusive saccades in infants than in normal adults. This pattern was similar to adults with no vestibular function. The gain and number of saccades was higher at 2 months than at 4 months, perhaps reflecting increasing vestibular function with age. When the head was rotated and the body kept still (COR + VOR), the infant responses were the same as those obtained to just the VOR condition, while adult gain increased significantly. Reisman and Anderson suggest this change may be a reflection of an infant's increasing ability to integrate proprioceptive and vestibular information with age. However, in the lighted COR + VOR condition, the infants' performance was much better, suggesting that three sets of information (visual, proprioceptive and vestibular) are integrated better than two. Alternatively, it might suggest that the visual information was somehow driving or calibrating the other kinds of information.

Schor, Narayan and Westall (1983) studied optokinetic afternystagmus in 25 infants between the ages of 1 to 10 months using EOG. Infants sat inside a drum which was covered with a pattern of random dots which they viewed monocularly and which rotated horizontally at 25 deg/sec. When the drum stopped the lights were turned out. Infants less than 2 months of age showed a slow phase of optokinetic afternystagmus (OKAN), the aftereffect of OKN, in the nasalward direction, no matter what direction the dots had moved. By 4 to 5 months of age, the direction of the OKAN was dependent on, and matched, the stimulus direction. The authors suggest that the directional bias of OKAN in young infants may reflect immature or reduced direction selectivity in the afferent signals arising from the cortex.

Several studies have investigated the suppression of VOR and OKN in young subjects. Goodkin (1980) found little or no VOR suppression at 2 months, although it was almost adult-like at 3 months, Regal et al (1983) report VOR suppression in 1 and 3 month old infants, although, interestingly, the slow compensatory eye movements which served to cancel the head movements were sometimes interrupted by saccades. Meijler and Van Den Berg (1982) used OKN suppression to test contrast sensitivity by presenting 1 to 3 month olds with a drifting grating to induce OKN, then overlaid it with a stationary grating. If the infants stopped their OKN response, this was interpreted as detection of the stationary test grating. By varying the contrast of the test grating, Meijler and Van Den Berg found contrast sensitivities significantly better than earlier findings (Banks & Salapatek, 1978). While the details of the specific gratings used in this study were not presented, it is clear that OKN can be suppressed in young infants. Ornitz and Honrubia (1988) tested infants between 2 and 15 months of age to determine if they could suppress VOR with visual fixation during low frequency sinusoidal stimulation. Unfortunately the results were not broken down by age so there is no way to tell if these values changed. As a group, the infants showed a higher VOR gain than adults when a fixation target was available (.46 as compared to adult value of 0), and a higher VOR gain to sinusoidal motion (.89 vs. .41) and impulse motion (1.32 vs. .59) in the dark. Their time constant was longer than adults' to sinusoidal stimulation (11.4 vs. 18.1 sec), but essentially the same to impulse (10.8 vs. 9.9 sec).

In summary, the VOR is well developed in young infants, but not adult-like. VOR gain is much larger in infants, while the VOR's time constraints are shorter, or more intense, early in infancy. While the gain is still elevated throughout childhood and reaches adult levels only in adolescence, the VOR's time constants increase significantly in the first postnatal months and by 6-months they are essentially at adult levels.

E. Saccadic eye movements

Saccadic eye movements[6] are rapid conjugate eye movements used in scanning and localizing targets, or in the absence of targets. Humans and other foveate animals use saccades to center a target onto the fovea. The first saccade to a target usually falls a little short and the rest of the distance to the target is made up by either another small, corrective saccade, by a slow

6 Saccade - from the French saquer, to pull or to tug, especially small, quick jerks on a rein when riding or driving a horse.

corrective movement called a glissade, or by a combination of the two. Saccades are ballistic eye movements, meaning that once the signal to execute a saccade leaves the oculomotor nuclei the saccade cannot be changed or interrupted until the eye is in a new position. In general, the larger the saccade, the higher the velocity of the eye movement, with saccades of 400° /sec occurring in 10 deg movements and 700° /sec for 80 deg movements (Leigh & Zee, 1984). Most adult saccades are less than 15 deg, with larger eye movements being accompanied by head movements (Bahill, Adler & Stark, 1975). The latency to saccades in adults is about 200 msec (Leigh, & Zee, 1984). Subjects scanning a field frequently engage in a saccade followed by a saccade sequence. Prior to the "take off" of an elicited saccade, subjects must program the direction in which the eye is to move, and the magnitude of the distance to be covered by the saccade. These logical components of a saccade are straight-forward in description, but complicated in their internal processing of external stimuli, as well as their execution (see Abrams, Meyer & Kornblum, 1989; Sparks, 1989 and Aslin, 1989 for an examination of these parameters). There have been many investigations into saccades (see Becker, 1989 for recent overview and Westheimer, 1989 for an interesting history), and space dictates a condensation of topics, so the discussion here will concentrate mostly on elicited saccades, and not the scanning of targets (a few of the articles covering this important area: Haith, 1980; Bronson, 1982, 1990b; Hainline, 1981b; Hainline & Lemerise, 1985).

Saccades are also the fast components of OKN and VOR, though there is debate about whether those saccades and localization and scanning saccades share a common source or trigger location. One side argues that while these saccades are generated in response to different stimuli, the amplitude of a scanning saccade is highly correlated with the velocity of the head movement during the search which implies VOR involvement, while the other side argues that the quick phase of VOR or OKN uses "saccades in essence" (p. 14, Becker, 1989) and yields only an approximate position, while goal-directed saccades are highly accurate (Barnes & Prosser, 1981; Jurgens, Becker & Rieger, 1981).

A relatively subtle point concerns the conjugacy of saccades. While saccades are generally considered to be conjugate movements, researchers do not agree on this point and attribute the observed asynchronies to the measurement technique used (for review see Becker, 1989; Williams & Fender, 1977). Miyoshi et al. (1981) measured saccades in 10 adults using EOG. They found, on average, the abducting eye consistently moved 7.4 msec before the adducting eye on horizontal saccades. The adducting eye moved faster, however, and tended to overshoot more. This asymmetry was also found in OKN, but was not found in vertical saccades. Eye movements in the latter had the same latency to start, and the same velocity, though there was a tendency for the eyes to overshoot when saccading upward, though not down. Schaefer, Schonfelder and Lietz (1988) conducted a similar study with 3 to 6 year old children making 23° horizontal eye movements and found approximately a 10 msec difference in latency, with the abducting eye starting the saccade before the adducting eye 70% of the time.

Saccades in all directions are generated in the oculomotor region of the brainstem known as the paramedian pontine reticular formation (PPRF), also known as the "pontine gaze centre" (Nieuwenhuys et al., 1988). In addition, the rostral interstitial nucleus of the medial longitudinal fascicle (riMLF) serves as a premotor area for vertical saccades. Lesions in the PPRF eliminate horizontal saccades while lesions in the riMLF eliminate vertical saccades (Buttner & Buttner-Ennever, 1988). The SC and the frontal eye fields are involved in the visual orienting of the eye and head during saccades, but lesions to either location alone will stop saccades temporarily, but only lesions to both will produce a permanent deficit (Buttner & Buttner-Ennever, 1988; Goldberg & Segraves, 1989). As noted earlier, the cerebellar nuclei project extensively to the SC. These projections are organized into two pathways (May et al., 1990) and research indicates that they modulate the accuracy of the magnitude, direction and speed of saccades (Blanks, 1988; Keller, 1989, May et al., 1990) and correlate the spatial registration of the multi-modal *sensory* representations which are sent to the SC, with the *motor* map of saccade generation (May et al., 1990). The basal ganglia, substantia nigra and caudate nucleus contribute to the suppression and initiation of saccades by inhibiting or releasing input to SC (Hikosaka, 1989).

Several studies have measured prenatal eye movements using ultrasound technology. Birnholz (1981) found that the first detectable slow eye movements occurred around 16 weeks gestation and rapid eye movements at 23 weeks. Prechtl & Nijhuis (1983) also found slow eye movements around 16 weeks, but spotted more rapid movements at 18-20 weeks, with "episodes" of activity change taking place around this time.

Aslin and Salapatek (1975) measured saccades in infants under 2 months of age using EOG. When presented with a 2° target located either 10, 20, 30 or 40° in the periphery, these infants did not respond with a single large saccade and a smaller corrective one, as adults usually do. Rather they generated a series of saccades of approximately equal magnitude (hypometric saccades).

Thus, a target 10° from a center fixation point might elicit 3 saccades of approximately 3° each. This curious pattern wasn't due to random eye movements made by disinterested infants, because the youngest infants consistently saccaded in the correct direction of the target and disinterested infants would be saccading in many different directions. In addition, while the median latency of infant saccades was 500-800 msec, considerably slower than an average adult's 200 msec, some infants generated a directionally appropriate response in 240 msec, which is hardly a disinterested response. Nor is this pattern due to an inability of infants to generate large saccades, since they do so when in dark surroundings. One possible explanation might be that the optics of the infants' eyes degraded their image of the stimulus, making it harder to see and, thus, harder to locate, but while this might support multiple saccades to a peripheral target, it would not support the hypometric pattern (Aslin, 1987a). The spatial coordinate system may be inaccurate, so peripherally distant targets may be poorly mapped from the retina to higher centers. These areas are still immature, possibly causing sensorimotor control problems (Aslin, 1987a, 1989). The hypometric saccades may be the result of conservative action on the part of the motor component of the visual system. It may be that while infants can generate large saccades, only small saccades can be well-controlled and the visual system uses feedback after each saccade to program the next step. Infants may have to learn the interactions between stimulus input, neural output and post-saccadic information and smaller, low velocity saccades might help in the processing of the optic flow information (Aslin, 1989). Salapatek et al. (1980) found that once started on a saccadic series, young infants would continue this series if the target disappeared. This continuation could be due to slow feedback about the target location which did not terminate the saccadic series immediately, or it could be that the infants were just searching for the target along the same line they had originally seen it. Despite these numerous hypotheses, the source of these hypometric sacccades has not been found.

Roucoux et al. (1983) also found hypometric saccades in young infants. In general, the more eccentric the target the infant was localizing, the more saccades the young infant used, though the youngest infants tested, 5 weeks, would not localize targets 45 deg from their fixation point. These hypometric saccades usually were not accompanied by head movements. When head movements did occur, the sum of the eye movement and head movement usually overshot the target. While this pattern of saccades decreased in the first 4 months, adult-like patterns were not reported until 1 year of age.

Hainline (1984) and Hainline et al. (1984) investigated the magnitude-velocity relations of saccades after working out modifications and calibration procedures for the ASL CR system (Hainline, 1981b; Harris et al., 1981; Harris et al., 1984). As noted above, larger saccades are associated with higher velocities, and the main sequence maps out this relation (Bahill, Clark & Stark, 1975). Most of the data reported involve a scanning paradigm and infants between the ages of 14 and 151 days of age showed good saccadic velocities, on the order of 250-300 deg/sec, and did not show many examples of the hypometric saccades observed by Aslin and Salapatek (1975). The slopes of the infants' main sequences were not different between infant age groups, nor significantly different from adults. However, younger infants in particular showed frequent rapid oscillations and slower saccades (130 deg/sec) than found in adults when viewing simple geometric forms instead of complex patterns. The rapid oscillations had inter-movement intervals which, at around 100 msec, were shorter than average adult intersaccadic intervals (200 msec) and very different from the intersaccadic intervals of 500 msec reported by Aslin and Salapatek (1975). Some of them did approximate the shortest adult intersaccadic intervals, 50 msec, reported for adults in tasks involving elicited saccades and not scanning (Taumer, Mie & Kommerel, 1972). Since it was not possible to know where the infant was looking during the eye movements, nor where they were directing a series of eye movements, these rapid oscillations could have been due to attentional factors, as Hainline et al. suggest, or they could just be saccades peculiar to infants, as Hainline et al. concluded. A sheer speculation is that since there was no way to tell where the infants were directing these eye movements, nor where they intended to stop, it's possible that some of these oscillations were versions of oblique saccades which have been found in infants to have the x and y components executed in rapid sequence rather than simultaneously (Shea & Aslin, 1982, 1988). In adults the components overlap for most of the saccade, but do show an asymmetrical onset of the horizontal and vertical components (i.e., Smit, Van Opstal & Van Gisbergen, 1990; Smit & Van Gisbergen, 1990). While the slow saccades were not dissimilar to some shown by adults with cerebellar lesions, Hainline et al. stated that because these saccades were only present with certain stimuli, they may have been due to attentional factors. Whether they fell into the oscillation or saccade subcategory, these rapid eye movements were more frequently classified as horizontal (43% saccades, 66% oscillations) than oblique (40% and 30%) or vertical (17% and 4%).[7]

[7] No definition is given as to how broad these categories were, so it's not clear, for example, whether a saccade 10 deg from horizontal was classified as horizontal or oblique. It would have been interesting to see the infant's main sequences broken down by

In a different paradigm, Lynch and Aserinsky (1986) used EOG to measure eye movements in sleeping and awake infants who were 4, 8 and 12 weeks of age. They concurred with other researchers that the percentage of conjugate eye movements increased with age. In agreement with Hainline et al. (1984), they also reported more horizontal eye movements (eye movements within +/-12° of the horizontal axis) than oblique, and more oblique than vertical (eye movements within +/-12° of the vertical axis).

In a finding similar to the smooth pursuit tracking data that show infants of around 3 months of age generating predictive eye movements to oscillating targets (Aslin, 1981), Haith, Hazan and Goodman (1988) reported that 3.5 month old infants who were presented with a series of visual events would begin to form expectations for events within the series and reliably execute saccades in anticipation of the next event.

Like the findings discussed for smooth pursuit and VOR, young infants show a significant improvement in their saccadic eye movements while only a few months old, but their saccades continue to change after infancy. The long saccadic latency of young infants, for example, tapers off through childhood, and possibly early adolescence before consistently attaining the adult latency (Cohen & Ross, 1978; Groll & Ross, 1982; Ross & Ross, 1983, 1987). Some of these changes are undoubtedly attentional factors, but it should be noted that the patterns of saccadic latencies, i.e., in response to predictable target displacements, is qualitatively the same as adults (Ross & Ross, 1987), showing similar changes to stimulus alterations, and even similar variability.

5. Speculation and conclusions

Given the huge literature and vast database which has built up around eye movements in the past several decades, it is not surprising that a number of models have been proposed to help explain the types of eye movements in their normal patterns, their developmental patterns, and their atypical patterns after lesions or cortical or muscular damage. Since models of eye movements are covered elsewhere, they will not be covered here (but the reader is referred to *recent* models: Abrams, et al., 1989; Aslin, 1989; Buttner & Buttner-Ennever, 1988; Chubb & Fuchs, 1981; Eckmiller, 1987; Huebner, Saidel & Leigh, 1990; Ron, Schmid & Orpaz, 1989; van Gisbergen & van Opstal, 1989; Westine & Enderie, 1990). One daunting thing about models, however: they don't even agree on how many kinds of eye movements there are. Tradition holds that there are five types of eye movements (or four, if one puts OKN under other categories) which include, basically, one fast type (saccades) and three relatively independent slow

direction of eye movement since this is correlated with variation in adult main sequence parameters (see Smit et al., 1990 for summary) possibly due to the vergence involved in downward-going saccades but not upward (Enright, 1989).

systems (smooth pursuit, vestibular, vergence) (Buttner & Buttner-Ennever, 1988). Each of these eye movements would respond to different sensory signals and would be controlled by separate, though linked, neural pathways. These pathways would merge at the motoneuron of the eye muscles. A separate argument has been made recently, however, which argues for two subsystems of eye movements - one fast, the saccadic subsystem, and one slow, the smooth subsystem (Bouis & Vossius, 1983; Collewijn et al., 1990; Steinman, Kowler & Collewijn, 1990).

In addition to the inherent appeal of any application of Occam's Razor (Morgan, 1894) which calls for simplifying things whenever possible, considerable logic supports this argument. The premotor activity of many of the five types of eye movements does overlap (Nieuwenhuys et al., 1988; Steinman et al., 1990), although one could argue this is just another example of natural redundancy in the system. In addition, just because the same area is active for different eye movements does not mean the pattern of activity is the same, let alone the eye movements resulting from it. Nevertheless, the ocular motoneurons receive input from two major structures: the PPRF-riMLF combination, and, second, the vestibular nuclei. It is clear that the former are involved when humans (and other organisms studied) generate horizontal and vertical saccades. The latter are involved "in every type of smooth eye movement" (Steinman et al., 1990). Given what we now know about the developmental aspects of eye movements, it does seem that there are distinct patterns to the development of rapid and slow eye movements and that the slow movements are more similar than they are distinct in some aspects, but considerable work remains to be done on these hypotheses.

Instead of dealing with models, this chapter will conclude with some speculations about developmental eye movement patterns. In addition to the dominant theme of describing the development of eye movements, with an eye toward the canon that babies get better as they get older, a second, more subtle and considerably more tenuous theme has emerged: namely that there exists a temporal pattern to the ability of infants to do things well which seems to be more than coincidental.

Eye movements change rapidly in young infants and they change in a familiar pattern from infant to infant. While there is considerable inter-individual variability in performance and one infant may begin to show reliable evidence of smooth pursuit at 9 weeks, and another not until 13 weeks, it is less obvious why the changes take place in this pattern, and what would "motivate" or "trigger" such a set of changes. There are a frustratingly large number of different sites in the visual system where these changes could be occurring: attentional, anatomical, physiological, sensory, or sensory-motor. It could be that the infants are not paying attention to the stimuli. It could be that the ocular muscles are not capable of generating certain types of eye movements at certain ages, i.e. smooth pursuit tracking at birth. It could be that the signals controlling the muscles are incomplete or inaccurate - the muscles may be capable of generating smooth pursuit but they are not receiving the signal from the oculomotor nuclei (or higher sites) to trigger such a movement. It could be that the infants are simply not seeing the targets well enough to track them or localize them, even though experimenters take great pains to ensure that contrast sensitivity and acuity thresholds are met. And it could, of course, be one of the greater frustrations: a combination of all these factors, and/or something no one has yet defined.

A. Oculomotor and Vestibular interaction - a speculative coincidence

In studies where the young infants are held by adults (i.e. Aslin & Jackson, 1979; Aslin & Dobson, 1983; Hainline, 1981b; Hainline, 1984; Hainline et al., 1984; Shea & Aslin, 1990; etc.) one quickly finds that when the infants are quite young, i.e. 4 weeks, they can't hold their own heads up very well or for very long, and they settle easily into an adult holding their head. In fact, most infants this age snuggle into the supporting hand, and get so comfortable they eventually go to sleep. Infants this age can give a large number of trials in a test session, if one is lucky enough to have caught them between naps and feeding, but the eye movements they show are, not too surprisingly, immature. Around 6 to 14 weeks, the eye movement patterns of vergence or smooth pursuit, for example, change drastically as they shift from an "infant-like" to an "adult-like" pattern. Also around this time, however, experimenters notice that infants can now hold their heads up by themselves and are less "interested" in having their heads held for them. So while we'd like to get more data per infant, they become too "rowdy" to test for long periods of time. By "rowdy" I mean they pull their heads away or tense up and close their eyes after just a few trials. It is still possible to collect large amounts of data from these infants, partly because they stay awake for longer periods of time, but the data collection tends to be in spurts of 4 or 5 trials at a time with long inter-data breaks.

It is a frequently expressed wish that "if only the older infants with the 'better' eye movements couldn't hold their heads well! Then we could hold their heads and collect more data on the 'adult-like' eye movements". In addition, parents spontaneously report a change in the infant's "appearance" around this age and describe them as "alert" or "more personable" or that they feel that the baby is "less aloof, has more eye contact now that they hold their heads up". After some time, the question arises: "What if this isn't a coincidence? What if the infant's ability to hold their head up is functionally correlated with the rapid change in eye movements at this age? What if this is a sign of overall coordination of the musculature and the control of the musculature?" And what if, as was pointed out earlier, the temporal coincidence of the onset of "mature" vestibular responses and the onset of the righting reflex and the offset of "primitive brainstem reflexes" is not a coincidence (Eviatar & Eviatar, 1979)?

Consider the timetable for typical changes in motor control which occur before 5 months of age (see Table 2). While good supine control of the head and neck appears around the fifth postnatal month, infants are certainly capable of head control in some positions and in reaction to some stimuli before that age. Most of these stimuli are vestibular, but as Regal et al. (1983) point out, young infants can coordinate eye and head movements to localize visual targets. The infants' movements are more likely to consist of multiple eye and head movements, and initial latency of movements and time-to-target acquisition are longer, but the velocity and duration of the individual components of these movements are quite mature. The VOR exists in 1 month olds but rapid changes are taking place over the first few months. It should be clear from the pages above that the brainstem's vestibular nuclei receive significant information from multi-modal sources, as do the SC, the cerebellum, and the oculomotor nuclei. Some of the information the oculomotor nuclei receive is from the vestibular nuclei and the canals of the labyrinth themselves, and the PPRF and vestibular nuclei send considerable signals to each other. Because there's little doubt that the vestibular system is involved with eye movements, consider what this might mean to a young infant:

> "Under normal circumstances not only the eyes but also the head is involved in tracking moving objects. The (VOR), which normally drives the eyes in the direction opposite to the head movement, has to be suppressed under these conditions. A current, satisfactory hypothesis is that the central nervous system actually generates a smooth pursuit signal to cancel out the VOR (Lisberger & Fuchs, 1978)". (Buttner & Buttner-Ennever, 1988, p. 16-17).

Not only does the subject, in our case an inexperienced, immature infant, have to generate smooth pursuit when tracking a slowly moving object, but they have to be able to shut the VOR off! And very young infants, around 1 month old, do not seem to be able to suppress VOR very well (Goodkin, 1980), or when they do, it's interspersed with saccades (Regal et al., 1983). Ornitz and Honrubia (1988) elaborate:

> "The minimal visual-vestibular interaction effects in the infants and the well-known influence of visual experience on the VOR suggest that in infancy, visual experience is not yet sufficient to modulate VOR function. The larger amplitude of the fast component of the nystagmus and the failure of fixation suppression of the VOR might reflect this lack of visual experience and/or immaturity of connectivity involving cerebellar and brainstem mechanisms. These centers of vestibular modulation may not be mature enough to fine tune the VOR". (p. 39)

It is an advantage for the VOR and cerebellum to remain plastic through infancy and childhood, and even through life so that changes in the body are compensated for by motor learning and do not destroy the balance or the ability to move (Llinas & Pellionisz, 1986; Lisberger, 1988), but if the VOR cannot be suppressed, or if the cerebellum is still being "programmed" and/or connected, the immaturity of infant oculomotor responses is not too surprising.

Given what researchers have learned about the developing retina and cortex and cell growth and dendritic proliferation throughout the visual system, it seems that an oculomotor and vestibular interaction is not the only thing keeping infants from behaving like adults. It also seems that too many visual capabilities - vergence, smooth pursuit, binocular fixation, stereoacuity, stereopsis (local and global) - improve too rapidly at too similar an age for there not to be a "trigger", or an underlying structure or "connection" which has changed. Another possible interaction could be that the development of accurate visual fixation allows calibration of the vestibular system or further calibration of the visual/vestibular interactions. What seems to be needed, then, is a study of the interplay of these systems to explore what correlation, if any beyond anecdotal, exists between the rapidly developing vestibular system and the beginning of rapid changes in the oculomotor system.

B. Clinical Similarities

Another thought which occurs to experimenters of young infants, usually when inspecting the sometimes peculiar patterns of eye movements their infants have generated, is whether these patterns exist anywhere else, perhaps in adults with abnormal eye movements, or in non-humans. It has been well-documented, for example, that schizophrenics show oculomotor abnormalities, predominantly saccadic disorders. These disorders, however, do not reflect the infant patterns, although both groups do show increased latencies (Thaker et al., 1989). Other injuries and diseases also cause abnormal patterns of eye movements. Wennmo et al. (1981) used EOG to test adult patients with lesions in different parts of the brainstem and found that pontine lesions lowered eye movement velocities like saccades more than medullary lesions. However, while the smooth component of OKN was affected in this way, smooth pursuit velocity was impaired by lesions at either site. When smooth pursuit velocity was lowered, saccades sometimes appeared in tracking. Ranalli and Sharpe (1988) used magnetic search coil techniques to measure vertical eye and head movements in 7 adult patients with brainstem damage, specifically the medial longitudinal fasciculus which connects the vestibular, oculomotor, trochlear and accessory ocular motor nuclei. These patients typically show horizontal gaze defects, and normal vertical saccades, but Ranalli and Sharpe also found the up/down VOR showed reduced gain and abnormal phase lag and reduced gain in vertical smooth pursuit with head free or fixed. The latter was attributed to ineffective cancellation of the VOR which manifested itself with higher than normal head movements in the head free condition, and a low smooth pursuit gain with head fixed.

Oculomotor disorders are also frequently found in patients with Huntington's disease. These disorders take the form of long, low velocity saccades, increased latencies in making saccades, and saccadic intrusions into fixation and smooth pursuit. The slow saccades are no more hypometric than controls (Collewijn, et al., 1988), as opposed to the slower, but consistently hypometric saccades of infants (Aslin & Salapatek, 1975). Saccadic intrusions during pursuit were not unilateral, but occurred equally in both horizontal directions. Vertical pursuit tended to be worse than horizontal but the saccadic intrusion was symmetrical whether the subject was tracking up or down (Collewijn et al, 1988). Smooth pursuit may also have low gain, but, for the most part, the smooth pursuit capability, as well as the smooth components of OKN and the VOR, are generally unaffected (Leigh & Zee, 1984; Collewijn, et al., 1988). One anatomical effect of Huntington's is the loss of neurons in the corpus striatum. The striatum projects to the pars reticulata of the substantia nigra in the basal ganglia (Nieuwenhuys, et al., 1988), which is involved with eye movements, especially its inhibitory projection to the SC (Hikosaka, 1989). Other structures are also implicated in these patterns of slow saccades (Keller, 1989; Kennard, 1986).

Interestingly, the inaccurate, saccade-interrupted smooth pursuit of young infants is sometimes mirrored in pursuit shown by adults with various cortical lesions or paralyzed oculomotor muscles (Leigh & Zee, 1984; Leigh, 1984). In Fig. 4A, an adult with a weakness of the lateral rectus OD shows saccades when tracking in one direction, and pursuit in the other. In Fig. 4B, a subject with a "deep right-sided lesion that included the posterior limb of the internal capsule" but no motion detection deficit showed asymmetrical pursuit with saccadic tracking interspersed with low gain pursuit in one direction and higher gain smooth pursuit in the other (Leigh, 1989), not dissimilar to the infant tracing shown in Fig. 3B. The catch-up saccades of the saccadic tracking were accurate.

Similar asymmetric saccadic intrusions and low gain occur in some Alzheimer's patients (Kuskowski et al., 1989), and have been interpreted as being caused by degeneration of the oculomotor centers in the cortex. This degeneration could lead to attentional problems and/or inappropriate and disruptive disinhibition of the pursuit process (see also Hikosaka, 1989).[8] Similarities, or correlations, of course do not imply causation, and the leap between these impaired adult subjects and the normal developmental pattern of young infants is a huge one. Nevertheless, the similarities in tracking are intriguing. The cerebellum plays a considerable role in the generation and control of eye movements and an immaturity or deficiency here could seriously impair an infant's or an adult's ability to generate accurate smooth pursuit. Alzheimer's and other dementias have been referred to as a "widespread disconnection syndrome" (p. 523, Katzman, 1986) of cortical association fibers and relay neurons, among other sites. If one considers that cell bodies in the infant's LGN are increasing in size, and the cortex is developing its dendritic proliferation and synaptic connections, perhaps the leap is not so broad.

8 My awareness of this literature is, unfortunately, due to the death of my cousin Brian in 1990 from Alzheimer's at the age of 38. It is to his memory that I dedicate this line of thought.

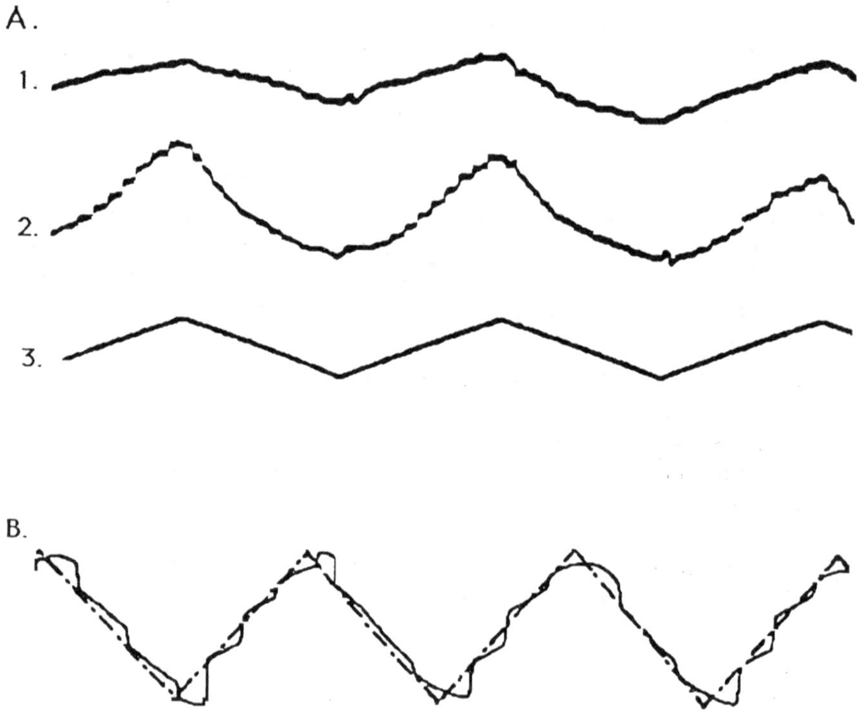

Fig. 4. Eye movement tracings of 2 adults to a smoothly moving target. (A) An adult with a right lateral rectus weakness viewing a target with the right eye. (1) is right eye position. (2) is left eye position. (3) is target position. Redrawn with permission from Leigh & Zee, 1984. (B) An adult with cerebellar disease tracking a target. Dashed line is target location. Redrawn with permission from Leigh, 1989.

C. Final thought

To finish where we started, with Henry Adams, we should recognize how far we have come out of our youthful understanding of the developmental aspects of eye movements, and take some comfort in the knowledge that, like the infants, their researchers know enough to know how to continue to learn.

References

Abramov, I., Gordon, J., Hendrickson, A., Hainline, L., Dobson, V. & La Bossiere, E. (1982). The retina of the newborn infant. *Science, 217*, 265-267.

Abrams, R.A., Meyer, D.E. & Kornblum, S. (1989). Speed and accuracy of saccadic eye movements: Characteristics of impulse variability in the oculomotor system. *Journal of Experimental Psychology: Human Perception and Performance, 15*, 529-543.

Aslin, R.N. (1977). Development of binocular fixation in human infants. *Journal of Experimental Child Psychology, 23*, 133-150.

Aslin, R.N. (1981). Development of smooth pursuit in human infants. In D.F. Fisher, R.A. Monty & J.W. Senders (Eds.), *Eye movements: Cognition and visual perception*. Hillsdale, N.J: Erlbaum.

Aslin, R.N. (1985). Oculomotor measures of visual development. In G. Gottlieb and N. Krasnegor (Eds.), *Measurement of audition and vision during the first year of postnatal life: A methodological overview*. Norwood, NJ: Ablex.

Aslin, R.N. (1986). Dark vergence in human infants: Implications for the development of binocular vision. *Acta Psychologica, 63*, 1-14.

Aslin, R.N. (1987). Motor aspects of visual development in infancy. In P. Salapatek and L.B. Cohen (Eds.), *Handbook of infant perception: From sensation to perception* (Vol. 1.), Orlando, FL: Academic Press, pp. 43-113. (a)

Aslin, R.N. (1987). Visual and auditory development in infancy. In J.D. Osofsky (ed.), *Handbook of infant development* 2nd edition. New York: Wiley, pp. 5-97. (b)

Aslin, R.N. (1987). Anatomical constraints on oculomotor development: Implications for infant perception. In A. Yonas (Ed.), *Perceptual development in infancy: The Minnesota symposia on child psychology*, Vol. 20. Hillsdale, NJ: Erlbaum, 67-104. (c)

Aslin, R.N. (June, 1989). Perception of visual direction in human infants. Paper presented at 23rd Carnegie-Mellon Symposium, "Visual perception and cognition in infancy". Pittsburgh, PA.

Aslin, R.N. (1992, in press). Infant accomodation and convergence. In K. Simons and D.L. Guyton (eds.). *Handbook of infant vision: Laboratory and clinical research*. New York: Pergamon Press.

Aslin, R.N. & Ciuffreda, K.J. (1983). Eye movements of preschool children: Comment. *Science, 222*, 74-75.

Aslin, R.N. & Dobson, V. (1983). Dark vergence and dark accomodation in human infants. *Vision Research, 23*, 1671-1678.

Aslin, R.N. & Jackson, R. (1979). Accommodative-convergence in young infants: Development of a synergistic sensory-motor system. *Canadian Journal of Psychology, 33*, 222-231.

Aslin, R.N. & Jackson, R. (1981). Dark vergence in human infants. *Investigative Ophthalmology and Visual Science, Supplement, 20*, 47.

Aslin, R.N. & Salapatek, P. (1975). Saccadic localization of visual targets by the very young human infant. *Perception & Psychophysics, 17*, 293-302.

Aslin, R.N. & Shea, S.L. (1987). The amplitude and angle of saccades to double-step target displacements. *Vision Research, 27*, 1925-1942.

Aslin, R.N. & Shea, S.L. (1990). Velocity thresholds in human infants: Implications for the perception of motion. *Developmental Psychology, 26*, 589-598.

Aslin, R.N., Shea, S.L. & Metz, H.S. (1990). Use of the Canon R-1 autorefractor to measure refractive errors and accommodative responses in infants. *Clinical Vision Sciences, 5*, 61-70.

Aslin, R.N. & Smith, L.B. (1988). Perceptual development. In M. Rosenzweig and L. Porter (Eds.). *Annual review of psychology*, Vol. 39. Palo Alto, CA: Annual Reviews, pp. 435-473.

Atkinson, J., Braddick, O., Wattam-Bell, J., Anker, S. & Norris, V.(1988). Videorefractive screening of accommodation performance in infants. *Investigative Ophthalmology & Visual Science, Supplement, 29*, 60 (15).

Bahill, A.T. & McDonald, J.D. (1983). Smooth pursuit eye movements in response to predictable target motions. *Vision Research, 23*, 1573-1583.

Bahill, A.T., Adler, D. & Stark, L. (1975). Most naturally occurring human saccades have magnitudes of 15 degrees or less. *Investigative Ophthalmology, 14*, 468-469.

Bahill, A.T., Clark, M.R. & Stark, L. (1975). The main sequence: a tool for studying human eye movements. *Mathematical Biosciences, 24*, 191-204.

Baker, R., Evinger, C. & McCrea, R.A. (1981). Some thoughts about the three neurons in the vestibular ocular reflex. In B. Cohen (ed.). *Vestibular and oculomotor physiology: International meeting of the Barany Society. Annals of the New York Academy of Sciences, 374*, 171-188.

Baloh, R.W., Sills, A.W., Kumley, W.E. & Honrubia, V. (1975). Quantitative measurement of saccade amplitude, duration and velocity. *Neurology, 25*, 1065-1070.

Baloh, R.W., Yee, R.D., Honrubia, V. & Jacobson, K. (1988). A comparison of the dynamics of horizontal and vertical smooth pursuit in normal human subjects. *Aviation, Space and Environmental Medicine, 59*, 121-124.

Banks, M.S. (1980). The development of visual accommodation during early infancy. *Child Development, 51*, 646-666. (a)

Banks, M.S. (1980). Infant refraction and accommodation. *International Ophthalmology Clinics, 20*, 205-232. (b)

Banks, M.S. & Salapatek, P. (1978). Acuity and contrast sensitivity in 1, 2 and 3 month old infants. *Investigative Ophthalmology & Visual Science, 17*, 361-365.

Banks, M.S. & Salapatek, P. (1983). Infant visual perception. In M.M. Haith and J.J. Campos (Volume eds.). *Handbook of child psychology: Infancy and developmental psychobiology*, (Vol. 2), (pp. 435-571). NY: Wiley, 4th edition.

Barbeito, R., Tam, W.J. & Ono, H. (1986). Two factors affecting saccadic amplitude during vergence: The location of the cyclopean eye and a left-right bias. *Ophthalmic and Physiological Optics, 6*, 201-205.

Barnes, G.R. & Prosser, A.J. (1981). Vestibular influence upon head-eye coordination. In B. Cohen (ed.). *Vestibular and oculomotor physiology: International meeting of the Barany Society. Annals of the New York Academy of Sciences, 374*, 560-570.

Becker, W. (1989). Metrics. In R. Wurtz & M. Goldberg, *The neurobiology of saccadic eye movements*. NY: Elsevier.

Becker, W. & Fuchs, A.F. (1985). Prediction in the oculomotor system: Smooth pursuit during transient disappearance of a target. *Experimental Brain Research, 57*, 562-575.

Bender, M.B. (1980). Brain control of conjugate horizontal and vertical eye movements: A survey of the structural and functional correlates. *Brain, 103*, 23-69.

Bergenius, J. (1984). Computerized analysis of voluntary eye movements: A clinical method for evaluation of smooth pursuit and saccades in oto-neurological diagnosis. *Acta-Oto-Laryngologica, 98*, 490-500.

Birch, E. & Held, R. (1983). The development of binocular summation in human infants. *Investigative Ophthalmology & Visual Science, 12*, 1103-1106.

Birnholz, J.C. (1981). The development of human fetal eye movement patterns. *Science, 213*, 679-681.

Blanks, R.H.I. (1988). Cerebellum. In J.A. Buttner-Ennever (ed.). *Neuroanatomy of the oculomotor system*. NY: Elsevier.

Boothe, R., Dobson, V. & Teller, D.Y. (1985). Postnatal development of vision in human and nonhuman primates. *Annual Review of Neuroscience, 8*, 495-546.

Bouis, D. & Vossius, G. (1983). Hierarchical organization of eye tracking movements. In R. Groner, C. Menz, D.F. Fisher and R.A. Monty (ed.). *Eye movements and psychological functions: International views*. Hillsdale, NJ: pp. 211-221.

Bradley, R. & Mistretta, C. (1975). Fetal sensory receptors. *Physiological Reviews, 55*, 352-382.

Braddick O., Atkinson, J., French, J. & Howland, H.C. (1979). A photorefractive study of infant accommodation. *Vision Research, 19*, 1319-1330.

Bronson, G.W. (1982). The scanning patterns of human infants: Implications for visual learning. *Monographs on Infancy, Whole No. 2.*

Bronson, G.W. (1983). Potential sources of error when applying a corneal reflex eye-monitoring technique to infant subjects. *Behaviour Research Methods & Instrumentation, 15*, 22-28.

Bronson, G.W. (1990). The accurate calibration of infants' scanning records. *Journal of Experimental Child Psychology, 49*, 79-100. (a)

Bronson, G.W. (1990). Changes in infants' visual scanning across the 2- to 14-week age period. *Journal of Experimental Child Psychology, 49*, 101-125. (b)

Brookman, K.E. (1983). Ocular accommodation in human infants. *American Journal of Optometry and Physiological Optics, 60*, 91-99.

Buttner-Ennever, J.A. (ed.) (1988). *Neuroanatomy of the oculomotor system*. NY: Elsevier.

Buttner, U. & Buttner-Ennever, J.A. (1988). Present concepts of oculomotor organization. In J.A. Buttner-Ennever (ed.). *Neuroanatomy of the oculomotor system*. NY: Elsevier.

Canning, C.R. & Restori, M. (1989). Doppler ultrasound velocity mapping of extra-ocular muscles: A preliminary report. *Eye, 3*, 409-414.

Chubb, M.C. & Fuchs, A.F. (1981). The role of the dentate nucleus y-group in the generation of vertical smooth eye movements. In B. Cohen (ed.). *Vestibular and oculomotor physiology: International meeting of the Barany Society. Annals of the New York Academy of Sciences, 374*, 446-454.

Cogan, D.G. & Kuwabara, T. (1986). Accessory cells in vessels of the paranatal human retina. *Archives of Ophthalmology, 104*, 747-752.

Cohen, M.E. & Ross, L.E. (1978). Latency and accuracy characteristics of saccades in children and adults. *Journal of Experimental Child Psychology, 26*, 517-527.

Collewijn, H., Steinman, R.M., Erkelens, C.J. & Regan, D. (1990). Binocular fusion, stereopsis and stereoacuity with a moving head. In D. Regan (ed.). *Vision and visual dysfunction, Vol. 10A: Binocular vision.* London: Macmillan.

Collewijn, H. & Tamminga, E.P. (1984). Human smooth and saccadic eye movements during voluntary pursuit of different target motions and different backgrounds. *Journal of Physiology, 351*, 217-250.

Collewijn, H., Went, L.N., Tamminga, E.P. & Vegter-Van der Vlis, M. (1988). Oculomotor defects in patients with Huntington's disease and their offspring. *Journal of Neurological Sciences, 86*, 307-320.

Conel, J.L. (1939). *The postnatal development of the human cerebral cortex.* Vol. 1: *The cortex of the newborn.* Cambridge, MA: Harvard University Press.

Conel, J.L. (1941). *The postnatal development of the human cerebral cortex.* Vol. 2: *The cortex of the 1-month infant.* Cambridge, MA: Harvard University Press.

Conel, J.L. (1947). *The postnatal development of the human cerebral cortex.* Vol. 3: *The cortex of the 3-month infant.* Cambridge, MA: Harvard University Press.

Crane, H.D. & Steele, C.M. (1985). Generation-V dual-Purkinje- image eyetracker. *Applied Optics, 24*, 527-537.

Dannemiller, J.L., Banks, M.S., Stephens, B.R. & Hartmann, E.E. (1983). Eye movements of preschool children: Comment. *Science, 222*, 75.

Dannemiller, J.L. & Freedland, R.L. (1989). The detection of slow stimulus motion in 2- to 5-month-olds. *Journal of Experimental Child Psychology, 47*, 337-355.

Dayton, G.O. & Jones, M.H. (1964). Analysis of characteristics of the fixation reflexes in infants by use of direct current electrooculography. *Neurology*, NY, *14*, 1152-1156.

Dayton, G.O., Jones, M.H., Aiu, P., Rawson, R.H., Steele, B. & Rose, M. (1964). Developmental study of coordinated eye movements in the human infant. I. Visual acuity in the newborn human: A study based on induced optokinetic nystagmus recorded by electrooculography. *Archives of Ophthalmology, 71*, 865-870. (a)

Dayton, G.O., Jones, M.H., Steele, B. & Rose, M. (1964). Developmental study of coordinated eye movements in the human infant. II. An electrooculographic study of the fixation reflex in the newborn. *Archives of Ophthalmology, 71*, 871-875. (b)

Dobbing, J. & Sands, J. (1973). Quantitative growth and development of human brain. *Archives of Disease in Childhood, 48*, 757-767.

Eckmiller, R. (1987). Neural control of pursuit eye movements. *Physiological Reviews, 67*, 797-857.

Enright, J.T. (1989). Convergence during human vertical saccades: probable causes and perceptual consequences. *Journal of Physiology, 410*, 45-65.

Eviatar, L. & Eviatar, A. (1979). The normal nystagmic response of infants to caloric and perrotatory stimulation. *The Laryngoscope, 89*, 1036-1045.

Eviatar, L., Eviatar, A. & Naray, I. (1974). Maturation and neurovestibular responses in infants. *Developmental Medicine and Child Neurology, 16*, 435-446.

Eviatar, L., Miranda, S., Eviatar, A., Freeman K. & Borkowski, M. (1979). Development of nystagmus in response to vestibular stimulation in infants. *Annals of Neurology, 5*, 508.

Evinger, C. (1988). Extraocular motor nuclei: Location, morphology and afferents. In J.A. Buttner-Ennever (ed.). *Neuroanatomy of the oculomotor system*. NY: Elsevier.

Fetter, M. & Buettner, U.W. (1990). Stimulus characteristics influence the gain of smooth pursuit eye movements in normal subjects. *Experimental Brain Research, 79*, 388-392.

Finocchio, D.V., Preston, K.L. & Fuchs, A.F. (1990). Obtaining a quantitative measure of eye movements in human infants: A method of calibrating the electrooculogram. *Vision Research, 30*, 1119-1128.

Finocchio, D.V., Preston, K.L. & Fuchs, A.F. (1991). Infant eye movements: Quantification of the vestibulo-ocular reflex and visual-vestibular interactions. *Vision Research, 31*, 1717-1730.

Fox, R., Aslin, R.N, Shea, S.L. & Dumais, S.T. (1980). Stereopsis in human infants. *Science, 207*, 323-324.

Freedland, R.L. and Dannemiller, J.L. (1987). Detection of stimulus motion in 5-month-old infants. *Journal of Experimental Psychology: Human Perception and Performance, 13*, 566-576.

Fuchs, A.F. (1971). The saccadic system. In P. Bach-y-Rita, C. Collins and J. Hyde, (eds.). *The control of eye movements*. NY: Academic Press. pp. 343-362.

Fulton, A.B., Dobson, V., Salem, D., Mar, C., Petersen, R.A. & Hansen, R.M. (1980). Cycloplegic refractions in infants and young children. *American Journal of Ophthalmology, 90*, 239-247.

Gallahue, D.L. (1989). *Understanding motor development: Infants, children, adolescents*. Indianapolis, IN: Benchmark Press, Inc.

Gay, A.J., Newman, N.M., Keltner, J.L. & Stroud, M.H. (1974). *Eye movement disorders*. St Louis, MO: C.V. Mosby Co.

Gesell, A., Ilg, F. & Bullis, G. (1949). *Vision: Its development in infant and child*. NY: Hafner.

Gilbert, P.F. & Thach, W.T. (1977). Purkinje cell activity during motor learning. *Brain Research, 128*, 309-328.

Goldberg, M.E. & Segraves, M.A. (1989). The visual and frontal cortices. In R. Wurtz and M. Goldberg, (eds.). *The neurobiology of saccadic eye movements*. NY: Elsevier.

Goodkin, F. (1980). The development of mature patterns of head-eye coordination in the human infants. *Early Human Development, 4*, 373-386.

Grantyn, R. (1988). Gaze control through superior colliculus: Structure and function. In J.A. Buttner-Ennever (ed.). *Neuroanatomy of the oculomotor system*. NY: Elsevier.

Green, D.G., Powers, M.K. & Banks, M.S. (1980). Depth of focus, eyesize, and visual acuity. *Vision Research, 20*, 827-836.

Groll, S.L. & Ross, L.E. (1982). Saccadic eye movements of children and adults to double-step stimuli. *Developmental Psychology, 18*, 108-123.

Hain, T. C. & Zee, D.S. (1989). Vergence. *Bulletin Societe Belge Opthalmologie, 237*, 145-161.

Hainline, L. (1981). An automated eye movement system for use with human infants. *Behavior Research Methods & Instrumentation, 13*, 20-24. (a)

Hainline, L. (1981). Eye movements and form perception in human infants. In D.F. Fisher, R.A. Monty & J.W. Senders (eds.). *Eye movements: cognition and visual perception*. Hillsdale, NJ: Erlbaum. (b)

Hainline, L. (1984). Saccades in human infants. In N. Gale and F. Johnson (eds.). *Theoretical and applied aspects of eye movement research*. NY: Elsevier, pp. 273-280.

Hainline, L. & Lemerise, E. (1985). Corneal reflection eye-movement recording as a measure of infant pattern perception: What do we really know? *British Journal of Developmental Psychology, 3*, 229-242.

Hainline, L., Turkel, J., Abramov, I., Lemerise, E. & Harris, C.M. (1984). Characteristics of saccades in human infants. *Vision Research, 24*, 1771-1780.

Haith, M.M. (1969). Infrared television recording and measurement of ocular behavior in human infants. *American Psychologist, 24*, 279-282.

Haith, M.M. (1980). *Rules that babies look by*. Hillsdale, NJ: Erlbaum.

Haith, M.M., Hazan, C. & Goodman, G.S. (1988). Expectation and anticipation of dynamic visual events by 3.5-month-old babies. *Child Development, 59*, 467-469.

Hallett, P. (1986). Eye movements. In K. Boff, L. Kaufman & J. Thomas (ed.). *Handbook of perception and human performance* (Vol, 1), (pp. 10-1 - 10-112). NY: Wiley.

Hansen, R.M. & Fulton, A.B. (1989). Psychophysical estimates of ocular media density of human infants. *Vision Research, 29*, 687-690.

Harris, C.M., Abramov, I. & Hainline, L. (1984). Instrument considerations in measuring fast eye movements. *Behavior Research Methods, Instruments & Computer Programs, 16*, 341-350.

Harris, C.M., Hainline, L., Abramov, I. (1981). A method for calibrating an eye-monitoring system for use with human infants. *Behavior Research Methods & Instrumentation, 13*, 11-17.

Harris, C.M., Hainline, L., Abramov, I., Lemerise, E. & Camenzuli, C. (1988). The distribution of fixation durations in infants and naive adults. *Vision Research, 28*, 419-432.

Harris, P.L., Cassel. T.Z. & Bamborough, P. (1974). Tracking in young infants. *British Journal of Psychology, 65*, 345-349.

Haynes, H., White, B.L. & Held, R. (1965). Visual accommodation in human infants. *Science, 148*, 528-530.

Held, R., Birch, E. & Gwiazda, J. (1980). Stereoacuity of human infants. *Proceedings of the National Academy of Sciences*, USA, *77*, 5572-5574.

Hendrickson, A.E. & Yuodelis, C. (1984). The morphological development of the human fovea. *Ophthalmology, 91*, 603-612.

Hepp, K., Henn, W., Vilis, T. & Cohen B. (1989). Brainstem regions related to saccade generation. In R. Wurtz and M. Goldberg (eds.). *The neurobiology of saccadic eye movements*. NY: Elsevier.

Herman, R., Maulucci, R. & Stuyck, J. (1982). Development and plasticity of visual and vestibular generated eye movements. *Experimental Brain Research, 47*, 69-78.

Hershenson, M. (1964). Visual discrimination in the newborn. *Journal of Comparative and Physiological Psychology, 58*, 270-276.

Hickey, T.L. (1977). Postnatal development of the human lateral geniculate nucleus: Relationship to a critical period for the visual system. *Science, 198*, 836-838.

Hickey, T.L. & Guillery, R.W. (1979). Variability of laminar patterns in the human lateral geniculate nucleus. *Journal of Comparative Neurology, 183*, 221-246.

Hickey, T.L. & Peduzzi, J.D. (1987). Structure and development of the visual system. In P. Salapatek and L. Cohen, *Handbook of infant perception: From sensation to perception*, (Vol. 1). NY: Academic Press. pp. 1-42.

Highstein, S.M. & Reisine, H. (1981). The ascending tract of Deiters' and horizontal gaze. In B. Cohen (Ed.). *Vestibular and oculomotor physiology: International meeting of the Barany Society. Annals of the New York Academy of Sciences, 374*, 102-111.

Hikosaka, O. (1989). Role of basal ganglia in saccades. *Revue Neurologique, 145*, 580-586.

Hikosaka, O. & Wurtz, R.H. (1989). The basal ganglia. In R.H. Wurtz and M. Goldberg (Eds.). *The neurobiology of saccadic eye movements*. NY: Elsevier.

Hitchcock, P.F. & Hickey, T.L. (1980). Prenatal development of the human lateral geniculate nucleus. *Journal of Comparative Neurology, 194*, 395-411.

Hollyfield, J.G., Frederick, J.M. & Rayborn, M.E. (1983). Neurotransmitter properties in the newborn human retina. *Investigative Ophthalmology and Visual Science, 24*, 893-897.

Howland, H.C. (1982). Infant eyes: Optics and accommodation. *Current Eye Research, 2*, 217-224.

Howland, H.C., Atkinson, J., Braddick O. & French, J. (1978). Infant astigmatism measured by photorefraction. *Science, 202*, 331-333.

Howland, H.C., Dobson, V. & Sayles, N. (1987). Accommodation in infants as measured by photorefraction. *Vision Research, 27*, 2141-2152.

Howland, H.C. & Sayles, N. (1983). Photorefractive studies of normal and handicapped infants and children. *Behavioral Brain Research, 10*, 81-85.

Howland, H.C. and Sayles, N. (1984). Photorefractive measurements of astigmatism in infants and young children. *Investigative Ophthalmology and Visual Science, 25*, 93-102.

Huebner, W.P., Saidel, G.M. & Leigh, R.J. (1990). Nonlinear parameter estimation applied to a model of smooth pursuit eye movements. *Biological Cybernetics, 62*, 265-273.

Isenberg, S.J. (1986). Macular development in the premature infant. *American Journal of Ophthalmology, 101*, 74-80.

Isenberg, S.J. (1989). General considerations of the newborn eye. In S.J. Isenberg (Ed.). *The Eye in infancy*. Chicago: Year Book Medical Publishers, Inc. pp. 31-47.

Isenberg, S.J., Molarte, A. & Vazquez, M. (1990). The fixed and dilated pupils of premature neonates. *American Journal of Ophthalmology, 110*, 168-171.

Juhola, M., Jantti, V. & Pyykko, I. (1985). Effect of sampling frequencies on computation of the maximum velocity of saccadic eye movements. *Biological Cybernetics, 53*, 67-72.

Jurgens, R., Becker, W. & Rieger, P. (1981). Different effects involved in the interaction of saccades and the vestibulo-ocular reflex. In B. Cohen (Ed.). *Vestibular and oculomotor physiology: International meeting of the Barany Society. Annals of the New York Academy of Sciences, 374*, 744-754.

Katzman, R. (1986). Alzheimer's disease. *Trends in Neuroscience, 9*, 522-525.

Keele, S.W. & Ivry, R. (1990). Does the cerebellum provide a common computation for diverse tasks? In A. Diamond (Ed.). *The development and neural bases of higher cognitive functions. Annals of the New York Academy of Science, 608*, 179-211.

Keller, E.L. (1988). The cerebellum. In R. Wurtz and M. Goldberg (Eds.). *The neurobiology of saccadic eye movements*. NY: Elsevier.

Keller, E. L. & Johnsen, S.D. (1990). Velocity prediction in corrective saccades during smooth-pursuit eye movements in monkey. *Experimental Brain Research, 80*, 525-531.

Kennard, C. (1986). Higher control mechanisms of saccadic eye movements. *Transactions of the Ophthamological Societies of the United Kingdom, 105*, 705-708.

Kowler, E. & Martins, A.J. (1982). Eye movements in preschool children. *Science, 215*, 997-999.

Kowler, E. & Martins, A.J. (1983). Eye movements of preschool children: Reply. *Science, 222*, 75-77.

Kremenitzer, J.P., Vaughan, H.G., Kurtzberg, D. & Dowling, K. (1979). Smooth pursuit eye movements in the newborn infant. *Child Development, 50*, 442-448.

Kuskowski, M.A., Malone, S.M., Mortimer, J.A. & Dysken, M.W. (1989). Smooth pursuit eye movements in dementia of Alzheimer's type. *Alzheimer's Disease and Associated Disorders, 3*, 157-171.

Larsen, J.S. (1971). The sagittal growth of the eye. I. Ultrasonic measurement of the depth of the anterior chamber from birth to puberty. *Acta Ophthalmologica, 49*, 239-262. (a)

Larsen, J.S. (1971). The sagittal growth of the eye. II. Ultrasonic measurement of the axial diameter of the lens and the anterior segment from birth to puberty. *Acta Ophthalmologica, 49*, 427-440. (b)

Larsen, J.S. (1971). The sagittal growth of the eye. III. Ultrasonic measurement of the posterior segment (axial length of the vitreous) from birth to puberty. *Acta Ophthalmologica, 49*, 441-453. (c)

Larsen, J.S. (1971). The sagittal growth of the eye. IV. Ultrasonic measurement of the axial length of the eye from birth to puberty. *Acta Ophthalmologica, 49*, 873-886. (d)

Larson, W. (1970). Clinical electro-oculography with suitable apparatus. *American Journal of Optometry, 47*, 295-303.

Lawrence, M.M. & Feind, C.R. (1953). Vestibular responses to rotation in the newborn infant. *Pediatrics, 12*, 300-305.

Leichnetz, G.R. & Goldberg, M.E. (1988). Higher centers concerned with eye movement and visual attention: cerebral cortex and thalamus. In J.A. Buttner-Ennever (Ed.). *Neuroanatomy of the Oculomotor System.* NY: Elsevier.

Leigh, R.J. (1989). The cortical control of ocular pursuit movements. *Revue Neurologique, 145*, 605-612.

Leigh, R.J. & Zee, D.S. (1984). *The neurology of eye movements.* Philadelphia: F.A. Davis.

Leuba, G. & Garey, L.J. (1987). Evolution of neuronal numerical density in the developing and aging human visual cortex. *Human Neurobiology, 6*, 11-18.

Ling, B. C. (1942). A genetic study of sustained visual fixation and associated behavior in the human infant from birth to six months. *Journal of Genetic Psychology, 61*, 227-277.

Lisberger, S.G. (1988). The neural basis for motor learning in the vestibulo-ocular reflex in monkeys. *Trends in Neuroscience, 11*, 147-152.

Lisberger, S.G. & Fuchs, A.F. (1978). Role of primate flocculus during rapid behavioral modification of vestibulocular reflex. I. Purkinje cell activity usually guided horizontal smooth pursuit eye movements and passive head rotation. *Journal of Neurophysiology, 41*, 733-763.

Llinas, R. & Pellionisz, A. (1985). Cerebellar function and the adaptive feature of the central nervous sytem. In A. Berthoz and G. Melvill Jones (Ed.). *Adaptive mechanisms in gaze control: Facts and theories.* NY: Elsevier.

Lynch, J.A. & Aserinsky, E. (1986). Developmental changes of oculomotor characteristics in infants when awake and in the 'active state of sleep'. *Behavioral Brain Research, 20*, 175-183.

Maunsell, J.H.R. & van Essen, D.C. (1983). The connections of the middle temporal visual area (MT) and their relationship to a cortical heirarchy in the macaque monkey. *Journal of Neuroscience, 3*, 2563-2486.

Maurer, D. (1975). The development of binocular convergence in infants. (Doctoral dissertation, University of Minnesota, 1974). *Dissertation Abstracts International B, 35*, 6136-B (University Microfilms No. 75-12, 121). (a)

Maurer, D. (1975). Infant visual perception: Methods of study. In L. Cohen & P. Salapatek (Eds.) *Infant perception: From sensation to cognition.* NY: Academic Press. pp. 1-76. (b)

May, P.J., Hartwich-Young, R., Nelson, J., Sparks, D.L. & Porter, J.D. (1990). Cerebellotectal pathways in the macaque: Implications for collicular generation of saccades. *Neuroscience, 36*, 305-324.

McBrien, N.A. & Barnes, D.A. (1984). A review and evaluation of theories of refractive error development. *Ophthalmic and Physiological Optics, 4*, 201-213.

McGinnis, J.M. (1930). Eye movements and optic nystagmus in early infancy. *Genetic Psychology Monographs, 8*, 321-430.

McGraw, M.B. (1941). Development of rotary-vestibular reactions of the human infant. *Child Development, 12*, 117-119.

Meijler, A. & Van den Berg, T. (1982). High contrast sensitivity in babies found using an eye movement reflex. *Documenta Ophthalmologica Proceedings Series, 31*, 229-235.

Metz, H.S. (1984). Calibration of saccades in infants. *Investigative Ophthalmology & Visual Science, 25*, 1233-1234.

Meyer, C.H., Lasker, A.G. & Robinson, D.A. (1985). The upper limit of smooth pursuit velocity. *Vision Research, 25*, 561-563.

Michaels, D.D. (1985). *Visual optics and refraction*. St Louis: C.V. Mosby. 3rd edition.

Michel, A.E. & Garey, L.J. (1984). The development of dendritic spines in the human visual cortex. *Human Neurobiology, 3*, 223-227.

Mitchell, T. & Cambon, K. (1969). Vestibular response in the neonate and infant. *Archives of Otolaryngology, 90*, 556-557.

Mitkin, A. & Orestova, E. (1988). Development of binocular vision in early ontogenesis. *Psychologische Beitrage, 30*, 65-74.

Miyoshi, T., Hiwatashi, S., Kishimoto, S. & Tamada, A. (1981). Dissociation of the eyes in saccadic movement. In B. Cohen (Ed.). *Vestibular and oculomotor physiology: International meeting of the Barany Society. Annals of the New York Academy of Sciences, 374*, 731-743.

Mohn, G. & van Hof-van Duin, J. (1986). Development of the binocular and monocular visual fields of human infants during the first year of life. *Clinical Vision Sciences, 1*, 51-64.

Mohindra, I., Held, R., Gwiazda, J. & Brill, S. (1978). Astigmatism in infants. *Science, 202*, 329-330.

Morgan, C.L. (1894). *Introduction to comparative psychology*. NY: Scribner's.

Munsinger, H. & Banks, M.S. (1974). Pupillometry as a measure of visual sensitivity among infants, young children, and adults. *Developmental Psychology, 10*, 677-682.

Mustarti, M.J. & Fuchs, A.F. (1989). Response properties of single units in the lateral terminal nucleus of the accessory optic system in the behaving primate. *Journal of Neurophysiology, 61*, 1207-1220.

Nieuwenhuys, R., Voogd, J. & Huijzen, C.V. (1988). *The human central nervous system*. 3rd edition. NY: Springer-Verlag.

Ornitz, E.M. (1983). Normal and pathological maturation of vestibular function in the human child. In R. Romand (Ed.). *Development of auditory and vestibular systems*. NY: Academic Press. pp. 479-536.

Ornitz, E.M., Atwell, C.W., Walter, D.D., Hartman, E.E. & Kaplan, A.R. (1979). The maturation of vestibular nystagmus in infancy and childhood. *Acta Otolaryngologica, 88*, 244-256.

Ornitz, E.M., Kaplan, A.R. & Westlake, J.R. (1985). Development of the vestibulo-ocular reflex from infancy to adulthood. *Acta Otolaryngologica, 100*, 180-193.

Ornitz, E.M. & Honrubia, V. (1988). Developmental modulation of vestibulo-oculomotor function. *Advances Oto-Rhino-Laryngology, 41*, 36-39.

Owens, D.A. & Leibowitz, H.W. (1980). Accommodation, convergence, and distance perception in low illumination. *American Journal of Optometry and Physiological Optics, 57*, 540-550.

Palmer, L.L. & MacDonald, J.E. (1990). Monocular and binocular measurements of smooth eye-pursuit by rural children in kindergarten and grade two. *Perceptual and Motor Skills, 70*, 608-610.

Peiper, A. (1963). *Cerebral function in infancy and childhood. NY: Consultants Bureau* (as cited in Yonas, A. and Owsley, C. (1987). Development of visual space perception. In P. Salapatek and L. Cohen (1987). *Handbook of Infant Perception: From perception to cognition*, (Vol. 2). NY: Academic Press. pp. 79-122).

Petrig, B., Julesz, B., Kropfl, W., Baumgartner, G. & Anliker, M. (1981). Development of stereopsis and cortical binocularity in human infants: Electrophysiological evidence. *Science, 218*, 1402-1405.

Precht, W. (1981). Visual-vestibular interactions in vestibular neurons: Functional pathway organization. In B. Cohen (Ed.). *Vestibular and oculomotor physiology: International meeting of the Barany Society. Annals of the New York Academy of Sciences, 374*, 230-248.

Prechtl, H.F.R. & Nijhuis, J.G. (1983). Eye movements in the human fetus and newborn. *Behavioral Brain Research, 10*, 119-124.

Provis, J.M. (1987). Patterns of cell death in the ganglion cell layer of the human fetal retina. *Journal of Comparative Neurology, 259*, 237-246.

Ranalli, P.J. & Sharpe, J.A. (1988). Vertical vestibulo-ocular reflex, smooth pursuit and eye-head tracking dysfunction in internuclear ophthalmoplegia. *Brain, 111*, 1299-1317.

Regal, D., Ashmead, D. & Salapatek, P. (1983). The coordination of eye and head movements during early infancy: A selective review. *Behavioral Brain Research, 10*, 125-132.

Reisman, J.E. & Anderson, J.H. (1989). Compensatory eye movements during head and body rotation in infants. *Brain Research, 484*, 119-129.

Rethy, I. (1969). Development of the simultaneous fixation from the divergent anatomic eye-position of the neonate. *Journal of Pediatric Ophthalmology, 6*, 92-96.

Riddell, P.M., Grose-Fifer, J., Hainline, L. & Abramov, I. (1991). Using photorefraction to assess the development of accommodation and convergence. *Investigative Ophthalmology & Visual Science, Supplement, 32*, 465.

Robinson, D. A. (1986). The systems approach to the oculomotor system. *Vision Research, 26*, 91-99.

Robinson, D. L. & McClurkin, J.W. (1989). The visual superior colliculus and pulvinar. In R. Wurtz and M. Goldberg (Eds.). *The neurobiology of saccadic eye movements*. NY: Elsevier.

Robinson, J. & Fielder, A.R. (1990). Pupillary diameter and reaction to light in preterm neonates. *Journal of Disease in Childhood, 65*, 35-38.

Ron, S., Schmid, R. & Orpaz, D. (1989). Applying a model of saccadic prediction to patients' saccadic eye movements. *Brain, Behavior & Evolution, 33*, 179-182.

Ross, S.M. & Ross, L.E. (1983). The effects of onset and offset warning and post-target stimuli on the saccade latency of children and adults. *Journal of Experimental Child Psychology, 36*, 340-355.

Ross, S.M. & Ross, L.E. (1987). Children's and adults' predictive saccades to square-wave targets. *Vision Research, 27*, 2177-2180.

Roucoux, A., Culee, C. & Roucoux, M. (1983). Development of fixation and pursuit eye movements in human infants. *Behavioral Brain Research, 10*, 133-139.

Salapatek, P., Aslin, R.N., Simonson, J. & Pulos, E. (1980). Infant saccadic eye movements to visible and previously visible targets. *Child Development, 51*, 1090-1094.

Salapatek, P. & Banks, M.S. (1978). Infant sensory assessment: Vision. In F.D. Minifie & L.L. Lloyd (Eds.). *Communicative and cognitive abilities: Early behavioral assessment*. Baltimore: University Park Press.

Salapatek, P., Bechtold, A.G. & Bergman, J. (1977, November). *Pupillary response in 1 and 2 month old infants*. Paper delivered to the Psychonomic Society, Washington, DC.

Salapatek, P. & Cohen, L. (1987a, b). *Handbook of infant perception*, (Vol 1,2). NY: Academic Press.

Salapatek, P., Haith, M.M., Maurer, D. & Kessen, W. (1972). Error in the corneal reflection technique: A note on Slater and Findlay. *Journal of Experimental Child Psychology, 14*, 493-497.

Salapatek, P. & Kessen, W. (1966). Visual scanning of triangles by the human newborn. *Journal of Experimental Child Psychology, 3*, 155-167.

Schaefer, A., Schonfelder, J. & Lietz, R. (1988). The interval of asynchronization during horizontal saccadic eye movements in young children. *Activitas Nervosa Superior, 30*, 193-195.

Schlag-Rey, M. & Schlag, J. (1989). The central thalamus. In R. Wurtz & M. Goldberg (Eds.). *The neurobiology of saccadic eye movements*. NY: Elsevier.

Schor, C.M., Narayan, V. & Westall, C. (1983). Postnatal development of optokinetic afternystagmus in humans. *Vision Research, 23*, 1643-1647.

Shea, S.L. (1988, April). *Infants' oculomotor responses to simultaneous and sequential visual target presentations.* Poster presented to the International Conference on Infant Studies, Washington, D.C.

Shea, S.L., & Aslin, R.N. (1984, April). *Horizontal, vertical and oblique eye movements in infants to single- and double-step target displacements.* Poster presented to the International Conference on Infant Studies, NY. (a)

Shea, S.L. & Aslin, R.N. (1984). Development of horizontal smooth pursuit in human infants. *Investigative Ophthalmology & Visual Science Supplement, 25*, 263. (b)

Shea, S.L. & Aslin, R.N. (1990). Oculomotor responses to step-ramp targets by young human infants. *Vision Research, 30*, 1077-1092.

Shea, S.L., Aslin, R.N., Fox, R. & Dumais, S.T. (1980). Assessment of stereopsis in human infants. *Investigative Ophthalmology & Visual Science, 19*, 1400-1404.

Shea, S.L., Doussard-Roosevelt, J.A. & Aslin, R.N. (1985). Pupillary measures of binocular luminance summation in infants and stereoblind adults. *Investigative Ophthalmology & Visual Science, 26*, 1064-1070.

Shea, S.L., Aslin, R.N. & McCulloch, D. (1987). Binocular VEP summation in infants and adults with abnormal binocular histories. *Investigative Ophthalmology & Visual Science, 28*, 356-365.

Sheena, D. & Borah, J. (1981). Compensation for some second order effects to improve eye position measurements. In D.F. Fisher, R.A. Monty & J.W. Senders (Eds.). *Eye movements: Cognition and visual perception*. Hillside, NJ: Erlbaum.

Shupert, C. & Fuchs, A. (1988). Development of conjugate human eye movements. *Vision Research, 28*, 585-596.

Simpson, J.I. & Graf, W. (1981). Eye-muscle geometry and compensatory eye movements in lateral-eyed and frontal-eyed animals. In B. Cohen (Ed.). *Vestibular and oculomotor physiology: International meeting of the Barany Society.Annals of the New York Academy of Sciences, 374*, 20-30.

Slater, A.M. & Findlay, J.M. (1972). The measurement of fixation position in the newborn baby. *Journal of Experimental Child Psychology, 14*, 349-364. (a)

Slater, A. & Findlay, J.M. (1972). The corneal reflection technique: A reply to Salapatek, Haith, Maurer and Kessen. *Journal of Experimental Child Psychology, 14*, 492-499. (b)

Slater, A.M. & Findlay, J.M. (1975). Binocular fixation in the newborn baby. *Journal of Experimental Child Psychology, 20*, 248-273. (a)

Slater, A.M. & Findlay, J.M. (1975). The corneal-reflection technique and the visual preference method: Sources of error. *Journal of Experimental Child Psychology, 20*, 240-247. (b)

Smit, A.C. & van Gisbergen, J.A.M. (1990). An analysis of curvature in fast and slow human saccades. *Experimental Brain Research, 81*, 335-345.

Smit, A.C., van Opstal, A.J. and van Gisbergen, J.A.M. (1990). Component stretching in fast and slow oblique saccades in the human. *Experimental Brain Research, 81*, 325-334.

Sokol, S., Moskowitz, A. & Paul, A. (1983). Evoked potential estimates of visual accommodation in infants. *Vision Research, 23*, 851-860.

Sparks, D.L. (1989). The neural encoding of the location of targets for saccadic eye movements. *Journal of Experimental Biology, 146*, 195-207.

Sparks, D.L. & Hartwich-Young, R. (1989). The deep layers of the superior colliculus. In R. Wurtz and M. Goldberg (Eds.). *The neurobiology of saccadic eye movements*. NY: Elsevier.

Spencer, R.F. & Porter, J.D. (1988). Structural organization of the extraocular muscles. In J.A. Buttner-Ennever (Eds.). *Neuroanatomy of the oculomotor system*. NY: Elsevier.

Steinman, R.M., Kowler, E. & Collewijn, H. (1990). New directions for oculomotor research. *Vision Research, 30*, 1845-1864.

Taumer, R., Mie, K. & Kommerell, G. (1972). Three kinds of reaction mechanisms of the human saccadic system. In H. Drischel (Ed.). *Biocybernetics IV*. Leipzig: VEB Gustav Fischer Verlag. pp. 236-242. (As cited in Becker, 1989).

Telang, G. & Dweck, H.S. (1989). The eye of the newborn: A neonatalogist's perspective. In S.J. Isenberg (Ed.). *The eye in infancy*. Chicago: Year Book Medical Publishers, Inc. pp. 3-8.

Teller, D.Y. & Movshon, J.A. (1986). Visual development. *Vision Research, 26*, 1483-1506.

Thaker, G., Kirkpatrick, B., Buchanan, R.W., Ellsberry, R., Lahti, A. & Tamminga, C. (1989). Oculomotor abnormalities and their clinical correlates in schizophrenia. *Psychopharmacology Bulletin, 25*, 491-497.

Torczynski, E. (1989). Normal development of the eye and orbit before birth: The development of the eye. In S. Isenberg (Ed.). *The eye in infancy*. Chicago: Year Book Medical Publishers, Inc. pp. 9-30.

Tronick, E. & Clanton, C. (1971). Infant looking patterns. *Vision Research, 11*, 1479-1486.

van Gisbergen, J.A.M. & van Opstal, A.J. (1989). Models. In R. Wurtz and M. Goldberg (Eds.). *The neurobiology of saccadic eye movements*. NY: Elsevier.

van Opstal, J. & van Gisbergen, J.A.M. (1990). Role of monkey superior colliculus in saccade averaging. *Experimental Brain Research, 79*, 143-149.

von Hofsten, C. (1976). The role of convergence in visual space perception. *Vision Research, 16*, 193-198.

Weale, R.A. (1990). Evolution, age and ocular focus. *Mechanisms of Ageing and Development, 53*, 85-89.

Weissman, B.M., Leigh, R.J. & Discenna, A.O. (1986). Maturation of the human vestibulo-ocular reflex during the first two months of life. *Developments in Oculomotor Research Satellite to the Cong. Int. Union of Physiol. Sci.*, Glenedon Beach. Oregon.

Weissman, B.M., Discenna, A.O. & Leigh, R.J. (1989). Maturation of the vestibulo-ocular reflex in normal infants during the first 2 months of life. *Neurology, 39*, 534-538.

Wennmo, C., Henriksson, N.G., Pyykko, I. & Schaleen, L. (1981). Eye-velocity programming in brain-stem disorders. In B. Cohen (Ed.). *Vestibular and oculomotor physiology: International meeting of the Barany Society. Annals of the New York Academy of Sciences, 374*, 774-783.

Westall, C.A. (1986). Binocular vision: Its influence on the development of visual and postural reflex eye movements. *Ophthalmic and Physiological Optics, 6*, 139-143.

Westheimer, G. (1989). History and methodology. In R. Wurtz & M. Goldberg. *The neurobiology of saccadic eye movements*. NY: Elsevier.

Westine, D.M. & Enderie, J.D. (1990). Temporal encoding for the control of saccades. *Biomedical Sciences Instrumentation, 26*, 175-180.

Wickelgren, L. (1967). Convergence in the human newborn. *Journal of Experimental Child Psychology, 5*, 74-85.

Wickelgren, L. (1969). The ocular response of human newborns to intermittent visual movement. *Journal of Experimental Child Psychology, 8*, 469-482.

Williams, R.A. & Fender, D.H. (1977). The synchrony of binocular saccadic eye movements. *Vision Research, 17*, 303-306.

Woestenburg, J.C., Verbaten, M.N. & Slangen, J.L. (1984). Eye- movements in a two-dimensional plane: A method for calibration and analysis using the vertical and horizontal EOG. *Biological Psychology, 18*, 149-160.

Wurtz, R.H. & Albano, J.E. (1980). Visual-motor function of the primate superior colliculus. *Annual Review of Neuroscience, 3*, 189-226.

Wurtz, R. & Goldberg, M. (Eds.) (1989). *The neurobiology of saccadic eye movements*. NY: Elsevier.

Young, L.R. & Sheena, D. (1975). Survey of eye movement recording methods. *Behavior Research Methods & Instrumentation, 1*, 397-429.

Yuodelis, C. & Hendrickson, A. (1986). A qualitative and quantitative analysis of the human fovea during development. *Vision Research, 26*, 847-855.

Zangemeister, W.H. & Stark, L. (1981). Active head rotations and eye-head coordination. In B. Cohen (Ed.). *Vestibular and oculomotor physiology: International meeting of the Barany Society. Annals of the New York Academy of Sciences, 374*, 540-559.

AUTHOR INDEX

O

P

SUBJECT INDEX

C

D

E

www.ingramcontent.com/pod-product-compliance
Lightning Source LLC
LaVergne TN
LVHW011909200926
848484LV00002B/577
9780444890054